Atmosphere Aerosol, Phytoplankton and its Influence on Climate Forming in the Pacific Ocean

Atmosphere Aerosol, Phytoplankton and its Influence on Climate Forming in the Pacific Ocean: Measurement New Methods

Edited by

Victoriya F. Yurchik

CAMBRIDGE SCHOLARS
PUBLISHING

Atmosphere Aerosol, Phytoplankton and its Influence on Climate Forming in the Pacific Ocean:
Measurement New Methods,
Edited by Victoriya F. Yurchik

This book first published 2011

Cambridge Scholars Publishing

12 Back Chapman Street, Newcastle upon Tyne, NE6 2XX, UK

British Library Cataloguing in Publication Data
A catalogue record for this book is available from the British Library

ISBN (10): 1-4438-2877-7, ISBN (13): 978-1-4438-2877-2

CONTENTS

EDITOR'S NOTE

Dear colleagues!

We are glad to present you with **Atmosphere Aerosol, Phytoplankton and Its influence on climate forming in the Pacific Ocean: New Measurement Methods**. It is a collection of new articles by young academics, students and PhD students, who participated in the First International Sailing University Conference on Climate Forcing, held on the board of the sailing training ship Nadezhda in the Sea of Japan and the Sea of Okhotsk at the end of summer 2010.

This collection was written by young researchers with encouragement and great support from their scientific advisors, and it presents a vivid overview of current problems in the research fields of atmosphere aerosol, phytoplankton communities, volcano activities and hydroacoustics, moreover it examines the influence of climate changes on phytoplankton communities.

This research work and the entire expedition would not have taken place without support from the Ministry of Education and Science of the Russian Federation and its contributions and grants to the young academics and scientific groups (State contracts: P163, 02.740.11.0439, 14.740.11.0139).

The idea and the entire concept of the Sailing University in the Far East of Russia as a unique combination of exciting on-board investigations and creative educational processes belongs to Dr. Oleg Bukin, Vice-Rector for Research of the Maritime State University. The results of his great contribution to the development of scientific programmes for young academics are present here.

Victoriya Yurchik
Director of the Department of Scientific and Educational Projects,
Maritime State University named after Admiral G.I. Nevelskoy

E-mail: vita_yurchik@mail.ru

PREFACE

This volume contains all papers presented on the board of the sailing ship Nadezhda by PhD students and researchers who participated in the annual scientific expedition of the International Sailing University 2010, organized by the Maritime State University named after Admiral G.I. Nevelskoy.

The Sailing University of the Maritime State University is intended to combine scientific experimental research methods with education processes, and to provide an international forum for young researchers and academic advisors.

Preliminary results were obtained in the Sea of Japan, the Sea of Okhotsk and in the Pacific Ocean within the period from August 3 till September 3, 2010. This collection presents a comprehensive coverage of the critical issues related to global climate changes, which were very noticeable in the seawater of the North-Eastern outlying areas of the Pacific Ocean during our latest expedition, where the Sailing University traditionally carries out its research activities.

This collection aims to elaborate on the research methods and on hardware used and tested throughout the expedition. Climate forcing and its influence on formation of phytoplankton communities was part of the expedition carried out in the Sea of Japan and the Sea of Okhotsk.

Dust storms in the continental areas of China and Mongolia, volcanic eruptions, and anthropogenic impact are of considerable influence on the ocean and atmosphere systems; while continental and volcano aerosol makes an impact on mineral nutrients for phytoplankton communities in seawater. The phytoplankton communities of the Sea of Okhotsk, where one of the largest bio recourses is concentrated, participate in the formation of atmospheric radioactive components, as this can be well observed in the area of Kashevarov Bank and Krusenstern Bank. Our scientific research aims to investigate the dynamics of the phytoplankton communities' behaviour under the impact of the climate forcing components typical of this region, i.e. tropic cyclones, dust storms, volcano aerosol emissions, and ozone layer changes.

On the other hand, the research intends to inspect how phytoplankton communities influence radioactive components in the near-water layer of the atmosphere (aerosol, carbon dioxide, methane) in order to determine the potential of this influence on photosynthetic activity of the phytoplankton cells, which in turn could possibly affect climate changes.

Students, young researchers and PhD students have prepared all the articles presented here, in cooperation with their academic advisors who also participated in the expedition and research activities. This way of training young researchers appears to be extremely effective, as all natural experiments are jointly carried out, and all lectures are devoted to the matter being under scientific investigation.

Dr. Oleg Bukin,
Vice-rector for Research, Professor
Maritime State University named after Admiral G. I. Nevelskoy;
Director of the International Sailing University

INTRODUCTION

OLEG A. BUKIN

Research results edited and collected in this book were obtained during the 10th scientific expedition of the International Sailing University on the board of the sailing ship Nadezhda in September 2010.

Since 1997, Maritime State University (Vladivostok, Russia) has been organizing scientific expeditions of the Sailing University. Mostly, research programs are focused on critical issues concerning the atmosphere and upper ocean layer monitoring, as well as the current state of phytoplankton communities. The Sailing University's goal is to research the main processes of the atmosphere and ocean that make contribution to the North-Eastern part of the Pacific Ocean climate formation; to determine how marine ecosystems react to those still ongoing climate changes; as well as to identify the role of phytoplankton communities in climate formation in the abovementioned area. Our key interests include potential use of phytoplankton communities for stabilization of sharp climate changes and investigation into their influence on the ocean carbon cycle. It is impossible to solve this problem without development of the new physical methods for atmosphere and ocean monitoring, therefore young researchers together with their academic advisors were assemble new equipment in order to apply new research methods, which are being tested in each scientific expedition. Consequently, the results of our former scientific expeditions as well as the Nadezhda's circum navigation (2003–2004) have become part of the Sailing University scientific programme. Articles written by Ekaterina Sokolova et al. and Maxim Shinkarenko present new equipment development in order to investigate the ocean upper

layer. The former work presents the development of Laser Induced Breakdown Spectroscopy (LIBS), which was implemented to investigate the elemental composition of seawater and phytoplankton cells. The above-mentioned method has been being developed for a long period of time, but in this case nanosecond pulse duration lasers were used in order to find out the quantity of the smallest element concentrations (10^{-3} g/l) in seawater and phytoplankton cells. Authors illustrate that LIBS lasers with femtosecond pulse duration usage increase test-sensitivity by two degrees, i.e. this method is extremely effective for investigation of the photosynthetic apparatus of phytoplankton cells, especially when the influence of various elements on the development rate of the phytoplankton communities is taken into account. Maxim Shinkarenko's work presents new equipment development in order to research phytoplankton concentration fields in the upper ocean layer: presented are testing results of the new pumpable two-frequency laser fluorimeter, which provides measurements of the seawater (phytoplankton included) fluorescence spectrums on the running vessel by excitation of two length waves of the laser radiation. This research made possible to elaborate on methodology in order to separate dissolved organic matter reproduced by fresh phytoplankton cells (young dissolved organic matter). Such methodology was created in order to appraise the effectiveness of the phytoplankton cells' photosynthetic apparatus during the production of organic matter. Konstantin Kluger's article presents processing methodology of the sea water fluorescence three-dimensional spectrum. The work of the PhD student Ivan Doroshenkov presents preliminary results of the fluorescence spectrums recorded throughout this expedition, aiming at studying dynamics of the research processes of degradation for the dissolved organic matter in seawater, as well as at defining the phytoplankton role in the reproduction processes of dissolved organic matter, and its significance in the carbonic cycle in the ocean.

The works of PhD students Igor Stepochkin, Vasily Kachur, and Victoriya Saklakova present the development of measurement algorithms for sea colour data. They also intend to develop methods for comparison

of satellite data and onboard measurements, and present the development results for semianalitic, empiric algorithms, and the other ones based on comparison of laser and solar induced fluorescence. The abovementioned works refer to the research of the effectiveness of satellite definition for the dissolved organic matter by the phytoplankton community. Presented is a detailed description of the data base, which collects important information and makes possible to process data on-line. Preliminary analysis of the atmosphere monitoring (satellite data, on-board photometric measurement and lidar) is presented in the article by the senior researcher Konstantin Shmirko. This analysis had been carried out within a period before Peak Sarychev eruption on the Kurile Islands, and after it (the eruption took place in June 2009, and was the biggest on the Kurile Islands for the last 300 years). The abovementioned research seeks to investigate how the volcanic eruption influenced the condition of phytoplankton communities in the Sea of Okhotsk. Another aspect of the interaction of phytoplankton communities with the atmosphere processes is presented in article by the student Alexey Bobrikov, who attempts to find out the experimental connection between atmosphere aerosol and phytoplankton concentrations in the lower underlayer of the atmosphere above the ocean. This work presents the results of the preliminary analysis, received during this expedition to the Kashevarov Bank area (the Sea of Okhotsk) and the Krusenstern Bank (the Pacific Ocean).

Andrey Storozhenko, a PhD student, describes the methods of acoustic measurements aimed at investigating the dynamics of the sonic scattering layer in the photic layer of the ocean, in order to observe and analyse the phytoplankton communities' dynamics and their reaction to various processes. The work by the PhD student Ilya Bukin, and the senior researcher Dmitry Lyakhov, analyses possible ways of using a remotely operating vehicle, equipped with the laser fluorimeter, in order to register seawater areas with high methane concentrations. Such works are to create a complex apparatus for research of the gas hydrates deposits and investigation of carbonic cycles in the ocean.

QUANTITATIVE ELEMENTAL DETERMINATION IN WATER BY FEMTOSECOND LASER-INDUCED BREAKDOWN SPECTROSCOPY

SERGEY GOLIK, ALEXEY ILYIN AND EKATERINA SOKOLOVA

Examination of the phytoplankton cell photosynthesis system condition and monitoring of concentration of sea water elements that limit phytoplankton cell development require an operative measurement method for determination of chemical elements concentration at the microgram-per-liter level. For this purpose, in our research we suggested and developed the method of nanosecond laser spectroscopy, that allowed for measurements of chemical elements in the liquid and phytoplankton cells at the $10^{-3} – 10^{-4}$ g/l level, which ensures its usability for real-time sea water and phytoplankton communities monitoring [1,2]. However, this method did not allow for the measurements of biogenous elements that limit phytoplankton cell development, and this prompted us to carry out a series of experiments aimed at determination of element concentration at the level of 10^{-6} g/l.

Research of elemental composition requires preliminary identification of spectral and molecular lines of plasma induced on the surface of sea water and distilled water. This research was carried out with an apparatus described in work [3].

Spectra of optical breakdown on the surface of water samples under normal atmospheric conditions were obtained for pulse duration – the *super-short* one, less than 50 fs, and the *subpicosecond* one of approximately 650 fs. Figure 1 shows the optical breakdown spectrum on the water surface induced with a laser pulse of 50 fs and registered without ICCD gating. This case reveals an intensive continuous spectrum resulting from a super-continuum generation, as well as Bremsstrahlung and recombination radiation. Figure 2 shows an averaging of 20 measurements of optical breakdown spectrum on the sea water surface induced with a laser pulse of 50 fs, registered with a time delay of 20 ns after laser pulse action. Laser pulsing was 10 Hz, exposure time of ICCD camera was 1 μs.

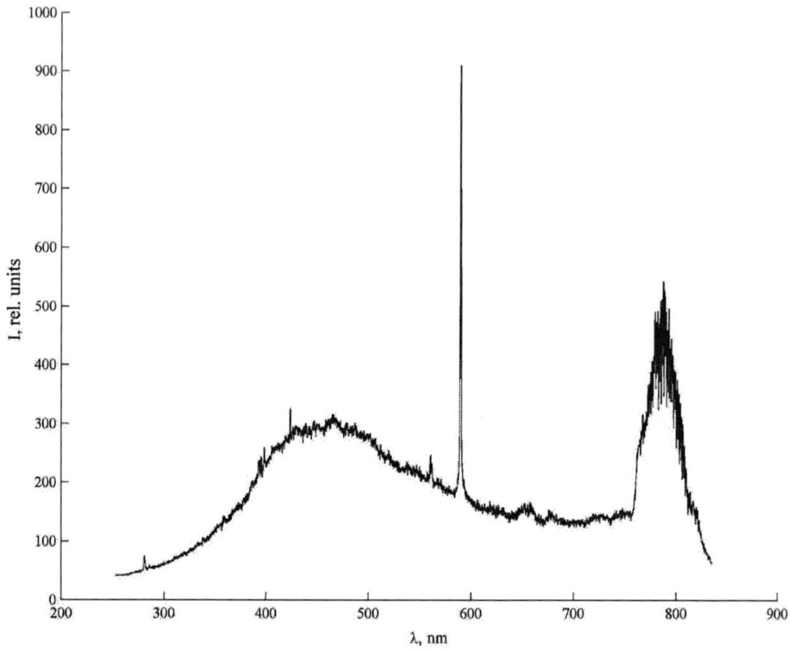

Fig. 1 – Spectrum of optical breakdowns on sea water surface with laser pulse duration of 50 fs, registered without gating.

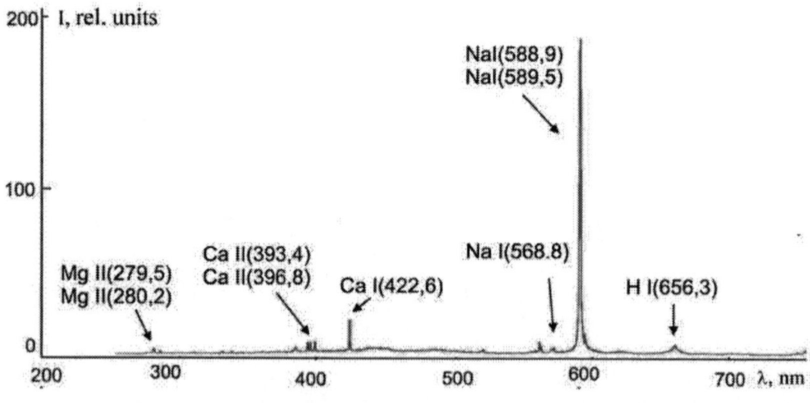

Fig. 2 – Spectrum of optical breakdown on sea water surface with laser pulse duration of 50 fs, time delay is 20 ns.

This case shows the absence of an intensive continuum. Emission lines of sea water macro-composition are registered. Let us note that all spectrums shown in this paper are not calibrated to the spectral sensitivity of recording equipment.

Optical breakdown spectra induced by pulses of 650 fs on the surface of sea and distilled water samples under normal atmospheric conditions, being an averaging of 20 laser pulses, are shown on figures 3 and 4 respectively. Within the given spectra were identified emission lines of sea water macro-composition (Ca, Mg, Na, K, H) as well as O and N lines. Figure 4 shows the distilled water spectrum induced by laser pulses of 650 fs under normal atmospheric conditions. H, N, and O lines were singled out in the distilled water spectrum. Displayed spectrums do not show lines of other elements in sea water, as their registration requires either special signal accumulation, or higher spectral resolution [3].

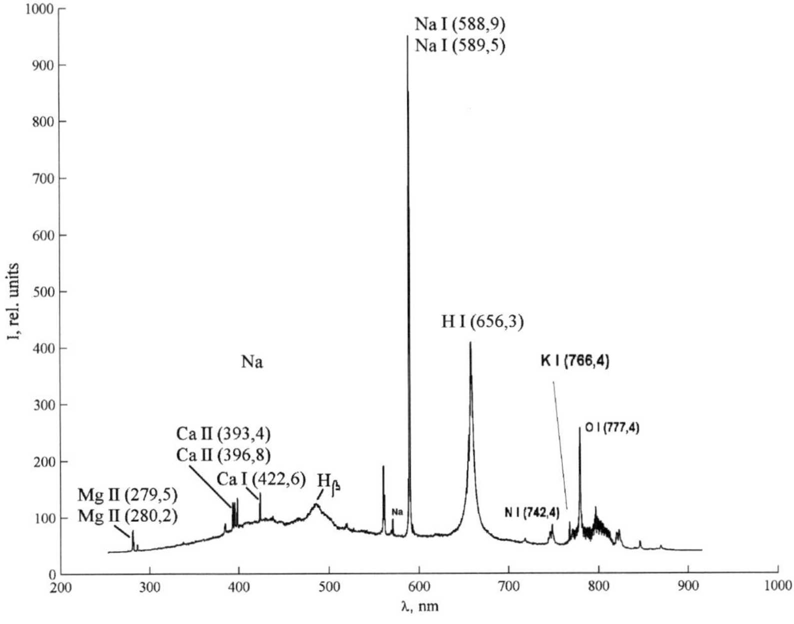

Fig. 3 – Optical breakdown spectrum on sea water surface, laser pulse duration is 650 fs, time delay is 40 ns.

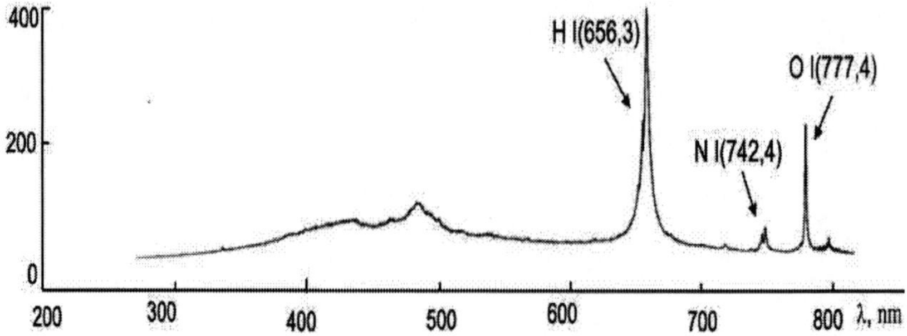

Fig. 4 – Optical breakdown spectrum on distilled water surface, laser pulse duration is 650 fs, time delay is 40 ns.

The experimental complex presented in work [3] was used for identification of spectral lines of phytoplankton cells in the femtosecond optical breakdown. Phytoplankton samples settled down on ash-free filters, so-called “white ribbon”, with an average pore size and high filtration speed. The volume of filtered sea water was 1 litre for each sample. Figure 5 shows laser plasma spectrums generated on the surfaces of studied filters with biological objects. For comparison, this paper shows research results of the two filtered sea water samples, one of them being one year old.

The spectrum was recorded with a time delay of 45±2,5 ns in relation to laser pulse, each 100 laser pulses being accumulated. At registration of the plasma spectrum within the wavelength from 560 to 600 nm and 740-780 nm, a neutral filter with 2.0 OD (Newport) was installed in front of the spectrograph slit; a neutral filter with 1.0 OD (Newport) was used for the rest of the range.

On the basis of spectral line table, data emission lines of the following elements were identified: K, Mg, Na, Ca, Cl, Fe, C_2, CH, O, H, N. Spectrums obtained showed relation of signal intensity (emission line intensity) to continuum intensity (SBR) on the scale not less than 10:1, also due to low intensity of the continuum, unlike that of the traditional nanosecond laser-induced breakdown spectroscopy. It is necessary to note that FeI emission lines under these conditions of excitation were not registered in sample #1 settled on the filter, which is presumably due to the decomposition of phytoplankton cells in sea water and with their passing through 10-micron filter pores. Sample #2, taken directly before its examination shows a high number of lines in the line spectrum of plasma, their intensity being higher than in sample #1 and in sea water.

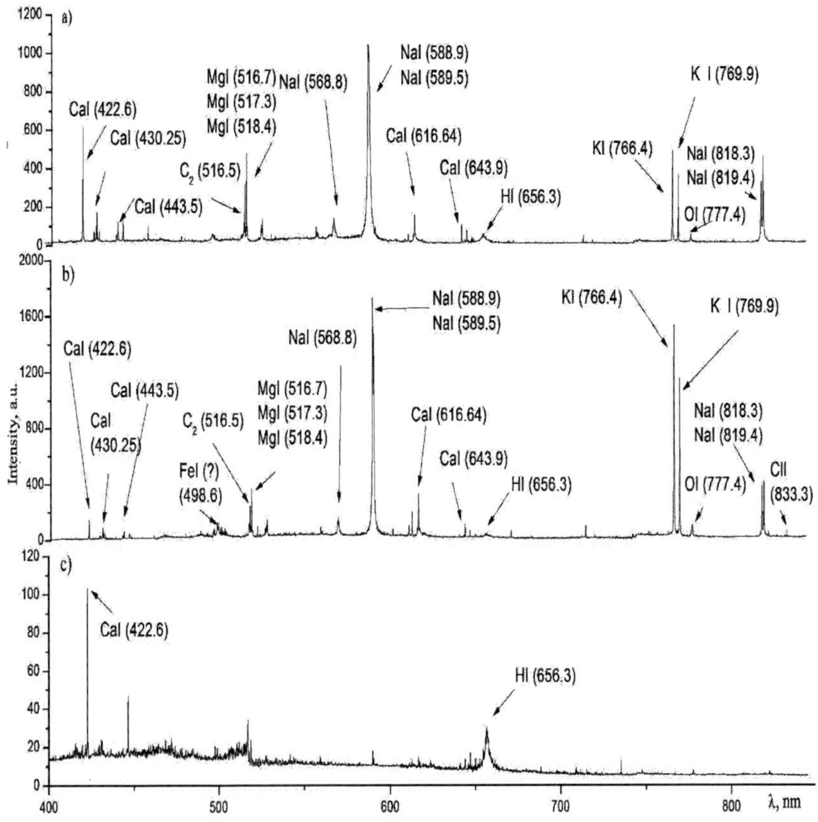

Fig. 5 - LIBS of filter samples: a) filter with sea water sample #1, b) filter with sea water sample #2, c) clean filter.

To analyze and assess minimally detectable concentrations of chemical elements in aqueous medium were used the State standard reference samples (SSRS) of water solutions of natrium ion water solutions (SSRS 7474-98) with the concentration of 0,95-1,05mg/cm^3 and calcium ion solutions (SSRS 7682-99). These elements were selected due to their high concentration in sea water and presence on LIBS spectrum of atomic and ion lines, which are clearly registered in cases of nano and femtosecond breakdown.

Dependence of Na I line intensity (588,9 nm) on concentration in the SSRS water solution helped research the limit of signal detection for Na.

Figure 6 shows averaging of 7 LIBS measurements of the SSRS Na spectrum. The spectrum was obtained with laser pulse of 100 Hz, and laser pulse duration of 650 fs. Energy pulse was 1.1 mJ, pulse duration was grope time delay of 75 ns, exposure time was 500 ns. The signal was accumulated on 800 laser pulses. Na emission line intensity on the wavelength of 588.9 nm was 40±8,5 relative units, the relative measurement error being 21% for this concentration. Thus, in defining the minimally detected concentration as a value of 3σ background signal, we obtained the Na atom detection limit: ~ 10^{-6} g/l [3].

Similarly, dependence line intensity of calcium II (393,4 nm) on concentration in the SSRS water solution helped research the limit of signal detection for calcium. The spectrum was obtained with laser pulsing of 1 kHz, pulse duration of 50 fs, energy per pulse of 1 mJ, and pulses time delay of 75 ns. Exposure time was 500 ns for each laser impulse, the signal was being accumulated during 4 seconds, and as a result 4000 laser pulses were summarized. Ca emission line intensity on the wavelength of 393,4 nm constituted (144±16) relative units, the relative measurement error for this calcium concentration in water being 11,1%. Thus, in defining the minimally detected concentration as a value of 3σ background signal, we obtain the Ca atom detection limit: ~ $2{\cdot}10^{-5}$ g/l.

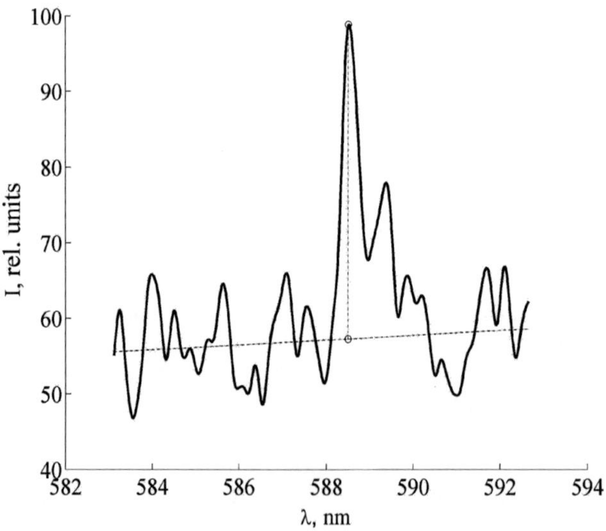

Fig. 6 - LIBS of NaCl water solution with Na concentration of $4{\cdot}10^{-6}$ g/l

Conducted experiments revealed that unlike in the case of nanosecond excitation, the detection limit for Ca and Na atoms with one-pulse femtosecond breakdown with the same analytical wavelength is two order magnitude weaker (see table 1). Further, the femtosecond breakdown increased sensitivity of spectrum definitions during analysis of vital biogenous elements, ferrum, for instance, the minimally detected concentration of which is far less reliable when nanosecond excitation is used.

Table 1. Comparison of minimally detected concentrations.

Element	Wave length, mn	Concentration, g/l (from work[4])	This work, g/l
Na I	588,9	$5\cdot10^{-4}$	10^{-6}
Ca II	393,4	$3\cdot10^{-4}$	$2\cdot10^{-5}$

Thus, firstly, the method of femtosecond laser-induced breakdown spectroscopy with signal time gating allows for registration of elemental composition of phytoplankton cells. Secondly, registration of microelements in samples deposited on filters is made possible. And, thirdly, this method confirms the necessity of carrying out the sea water and phytoplankton cell analysis directly after sampling.

The typical fluorescence time of femtosecond emission lines breakdown is considerably shorter compared with the nanosecond one. However, due to generation of stable femtosecond laser plasma on the surface of studied samples and the high ratio of line intensity to the continuum, the femtosecond LIBS showed high reliability and result reproducibility for concentration measurements of chemical elements in liquid.

References

Bukin O.A., Saliuk P.A., Mayor A.Yu., Golik S.S., Ilyin A.A., A.V.Bulanov, E.N.Baulo, D.A. Akmaikin. Use of Laser Spectroscopy Methods in Research of Carbon Cycle Elements in Ocean // Atmosphere and Ocean Optics, vol. 23, 2010, № 03, p.229-234.

Bukin O.A., Golik S.S., Ilyin A.A., Kulchin Yu.N., Sokolova E.B., Baulo E.N. Laser-Induced Spectroscopy of Fluid Mediums with Femtosecond-Length Impulse Induction. // Atmosphere and Ocean Optics, vol. 22, 2009, № 03, pp.296-300.

Bukin O.A., Aleskeev A.V., Ilyin A.A., Golik S.S., Tsarev V.I., Bodin N.S. Use of Femtosecond Laser-Induced Breakdown Spectroscopy with Multi-Impulse Plasma Induction for Monitoring Sea Water Quality and Phytoplankton Conditions // Atmosphere and Ocean Optics, vol. 16, 2003, № 01, p.26-32

Fiched P., Mauchien P., Wagner J-F., Moulin C.. Quantitative Elemental Determination in Water and Oil by Laser Induced Breakdown Spectroscopy // Analytica Chimica Acta.-2001.- v. 429.- №2.- p. 269-278.

OBSERVATION RESULTS OF AEROSOL DYNAMICS IN ATMOSPHERE OF SEA OF JAPAN AND SEA OF OKHOTSK

KONSTANTIN SHMIRKO

Recent times have seen close attention paid to studies of processes linked with climate changes. On a first-priority basis, this is relevant to dynamics of radioactively active atmosphere components, which include trace gases (ozone, CO_2, methane, water vapor, etc.), and atmospheric dust/volcanic/anthropogenic aerosol. An increasing number of volcanic eruptions and their increasing capacity during maximum value periods of volcanic activity cycles are thought to be the main cause of global temperature changes. This results in higher emissions of greenhouse gases and volcanogenic aerosols into atmosphere, which in turn brings about radiation balance changes.

Within a period from 2007 till present, the IACP FEB RAS lidar station has been monitoring volcanic aerosol distribution, caused by Okmok, Kasatochi and Sarychev Peak volcanoes. In all these cases, at the atmospheric tropopause level and lower, aerosol layers capable of changing the relation between flux difference of incoming and outgoing radiation at the tropopause level were observed, which results in troposphere temperature condition change. This influence directly depends on vertical distribution of scattering layers. The objective of this article is to highlight in what atmospheric changes the eruption of Sarychev Peak volcano on Matua Island resulted. We used data from shipborne lidar measurements together with spaceborne ones. In the latter case we used data from Cloud-Aerosol Lidar and Infrared Pathfinder Satellite Observation (CALIPSO) satellite provided by NASA: http://www-calipso.larc.nasa.gov/

Two scientific research expeditions were carried out in 2009 and 2010. The first voyage aboard the Senite towards the volcano took place two weeks after its eruption in July 2009. The second voyage took place a year after the first expedition, within a period from August 3 to September 2, 2010, aboard the training tall ship Nadehzda. The following atmosphere parameter measurements were continuously carried out aboard the ship throughout its entire route: aerosol optical depth, vertical distribution of backscattering and extinction coefficients, particle size distribution function as well as samplings for chemical testing.

A shopborne mobile lidar was used to obtain vertical profiles of the backscattering signal (Fig.1a). The receiving system diameter was 150 mm. Backward scattering signal registration was carried out with a 14-bit 10mhz AD converter. The transmission unit was based on the Nd:YAG Brilliant-Ultra laser (by Quantel) with pulse energy being equal to ~45 mJ. Laser impulse frequency was 15 Hz. Backscattering signal measurements were effectuated in clusters of 1000 impulses (in cases of altitudinal sounding), and of 100 impulses, when dynamics of light-scattering layers were being taken into account. The background signal measurements were taken for their further subtraction before each set of measurements.

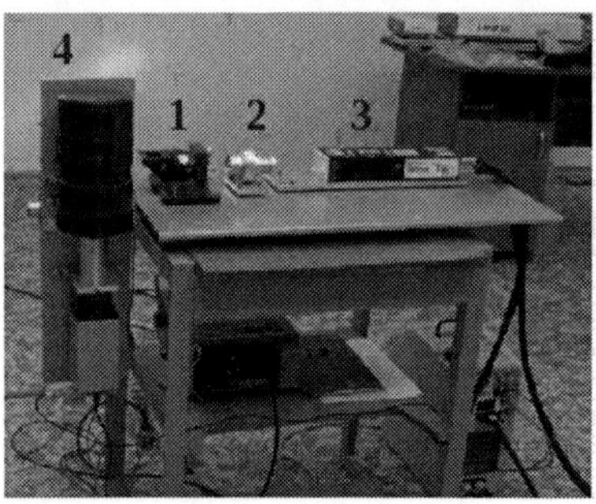

Fig.1a Shipborne mobile lidar:
1 sine V-block; 2 collimator; 3 laser head; 4 receiver.

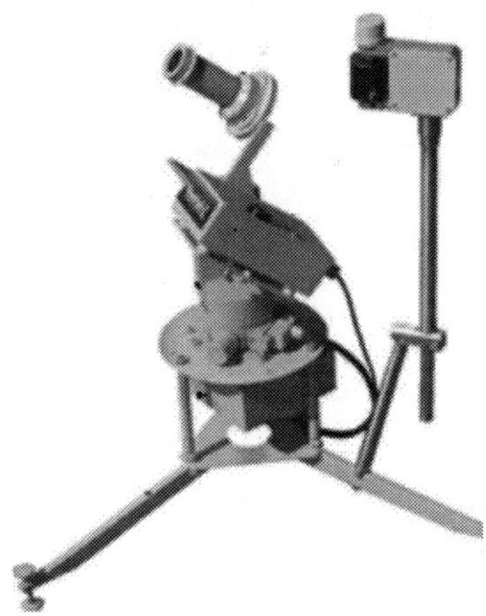

Fig.1b Sun photometer with sun tracking system.

Passive atmosphere sounding was carried out with a portable direct sun photometer (made in Tomsk) (Fig.1b), synchronously with lidar sounding, when possible. Measurements were taken hourly when the weather was clear (the lidar profile does not support nebulosity).

Apart from optical remote sounding, we carried out contact measurements of parameters of the atmospheric surface layer, which included measurements of particle size distribution function, atmospheric aerosols and black carbon concentration. Such measurements were carried automatically on an hourly basis.

Figure 2 shows expedition routes. On the map, green dots indicate lidar measurements, while yellow circles show spectrophotometric measurements.

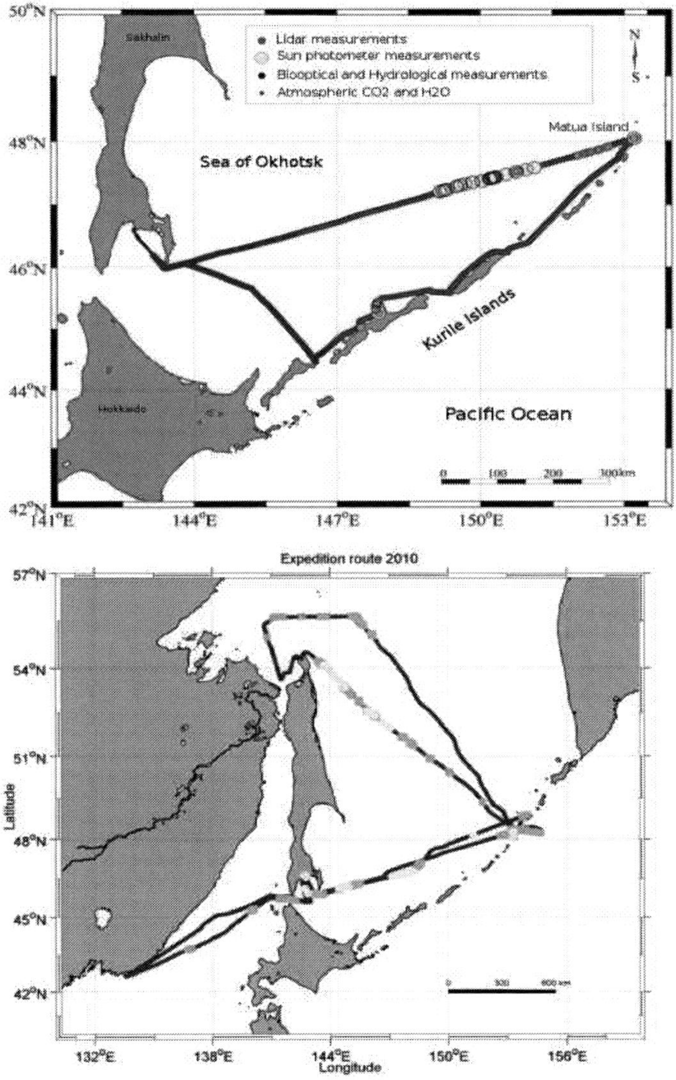

Fig.2 Routes a) 2009, b) 2010

Spatial resolution of satellite lidar data was 300 meters between profiles. However, their accumulation was necessary to obtain a high-quality signal of backward scattering. Being interested in the atmosphere condition in the vicinity of Matua Island, we were obtaining the lidar signal composite profile via averaging profiles within a circle of a 200 km radius.

Data obtained during expeditions can be conventionally divided into two types, belonging to either "pure" or "agitated" atmosphere. We define the "pure" atmosphere as a case when lidar profiles do not show strongly marked nebulosity, their integral coefficient of backward scattering being equal to, or less than 0.01 sr^{-1} m^{-1} Analogously, "agitated" atmosphere implies such situations when lidar profiles do not show strongly marked nebulosity, while its integral coefficient of backward scattering is bigger than 0.01 sr^{-1} m^{-1}. Values of backward scattering integral coefficient, which is used as a classification criterion, were obtained via statistic processing of lidar data of previous expeditions (2004, 2006, 2008).

Main (background) lidar profiles reflect the conditions typical of the Sea of Okhotsk and the Sea of Japan atmosphere:

- planetary boundary layer (PBL) lies within the range from 340m to 1000m (stable boundary layer lies within range from 170m to 340m);
- atmospheric aerosol is majorly concentrated in lower PBL levels;
- rather stable aerosol stratification is observed, its change being caused by meteorological processes.

After the eruption of Sarychev Peak volcano, a large amount of pyroclastics was emitted into the atmosphere. It reached the tropopause heights and formed a permanent aerosol layer there (Fig. 4a). Apart from considerable height from 8 to 12 km and high values of backscattering coefficient along the entire aerosol layer depth, the depolarization factor value equaled 0.2, which corresponds to non-spherical particles of dry origin (the volcanic one, in this case). This atmospheric condition was being observed throughout June-July 2009.

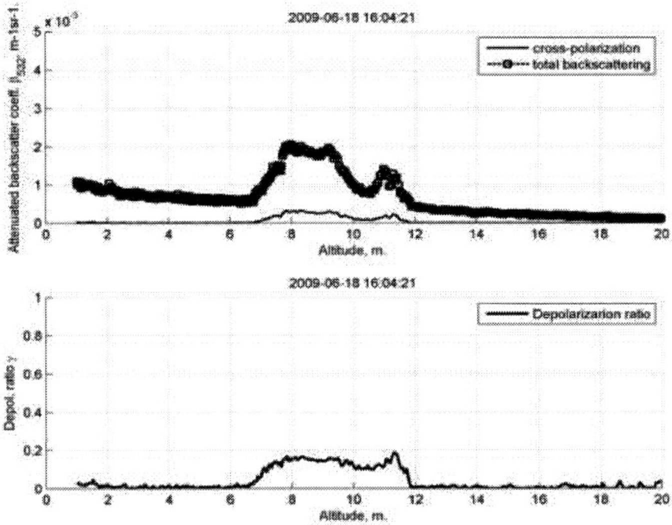

Fig.4a First successful shot produced by CALIPSO after volcanic eruption

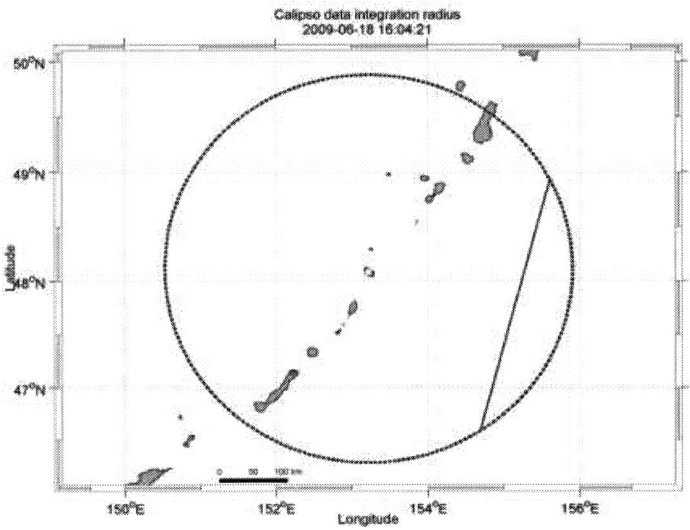

Fig. 4b Averaging area

Figure 5 shows lidar measurements taken during the expedition to the volcano in June 2009. The closer to the volcano, the more apparent the aerosol layer became at the height of 8-9.5 km, revealing the tendency of going even lower. Near the volcano itself, obtention of high-quality vertical profiles was not possible because of considerable ash emissions. A large amount of aerosol emitted into the atmosphere resulted in its stratification, and the planetary boundary level expanded its boundaries as aerosol was inflowing.

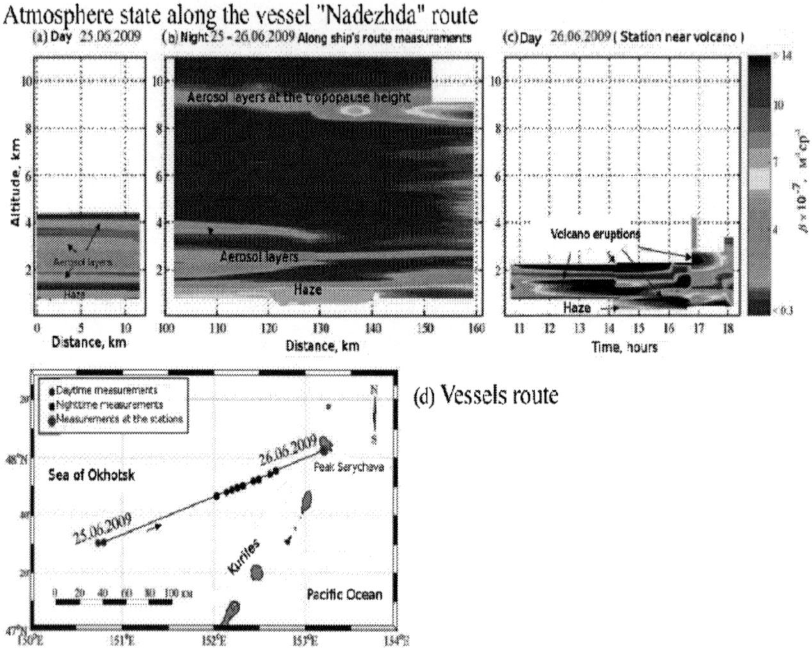

Fig. 5 Lidar experiment near Sarychev Peak volcano

The planetary boundary layer had a complex multi-layered structure. At that time we registered at least three sub-layers inside PBL of same origin – from ash and water vapor emissions. The first layer was located at 300 meters above the sealevel, the second and third had altitudes of 1500 and 2000 meters. The first 2 km were completely attenuated by atmosphere constituents, and we were unable to retrieve data from above 2 km. The planetary boundary layer at that time was located at the 2 km height. It should be noted that the marine boundary layer is located at the altitude less than 700-1100 m.

Due to the high values of aerosol concentration and optical thickness in the planetary boundary layer, there are possibilities in intensive gas exchange between atmosphere and the sea upper-top layer, as well as volcanic aerosol disposition onto the water surface.

Figure 6 shows CALIPSO satellite data we analyzed to explore the atmosphere structure at the higher altitudes.

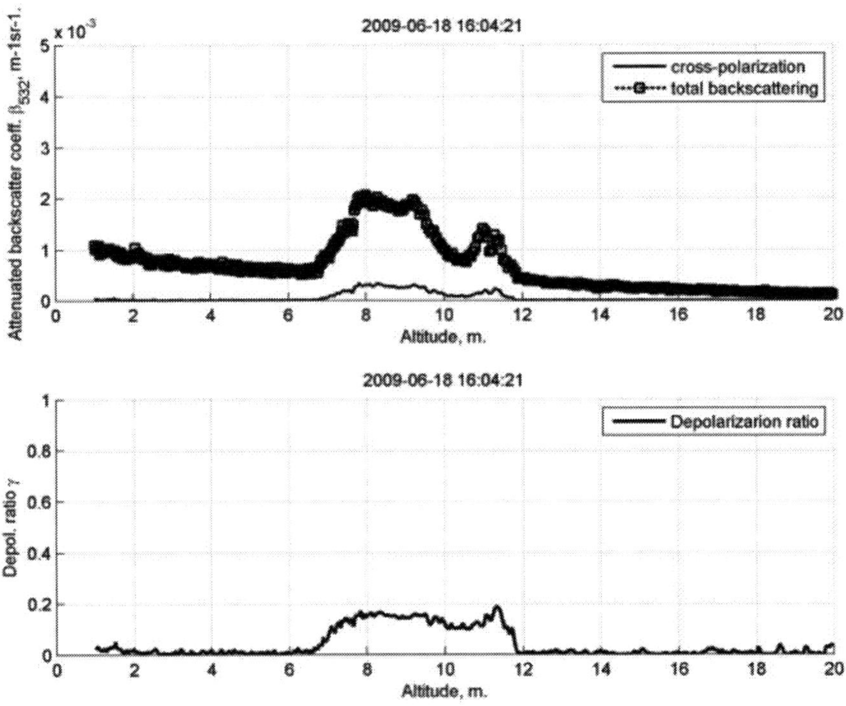

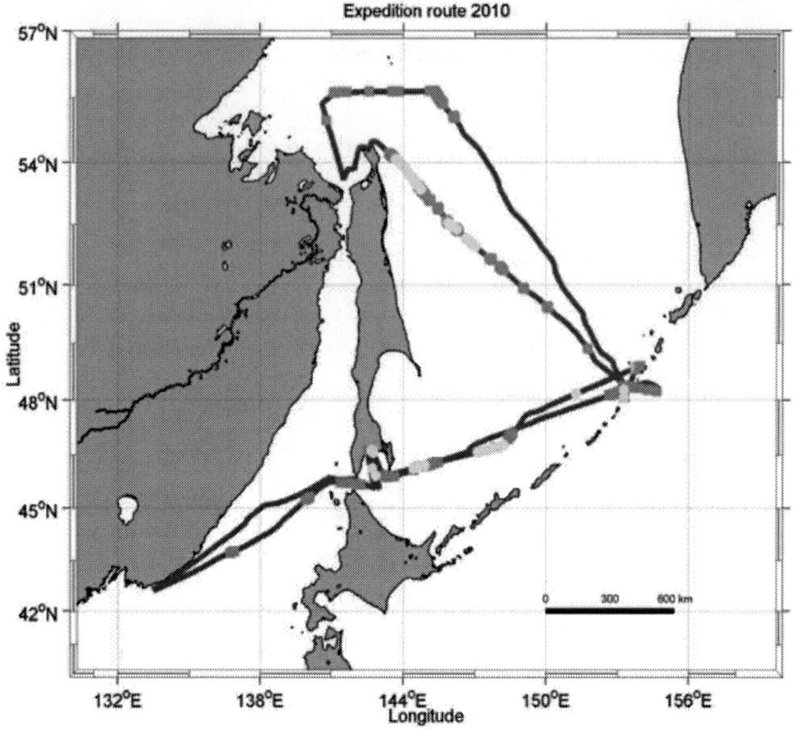

Fig.6 Vertical profile of backscattering signal (June 25, 2009) and integration range

That day CALIPSO satellite data clearly showed the aerosol layer position at the height of 8.5 km. Moreover, it was clear that the atmospheric bottom layer had been filled with aerosol up to the height of 2 km.

Expedition carried out in 2010 showed that the atmosphere in the vicinity of Matua Island had restored its intermediate qualities. Apart from the boundary layer and tropopause, aerosol traces were not present anymore. Residual aerosol could account for the layer at the tropopause height. At that time, the planetary boundary layer had no sub-layers and its altitude was 700 m.

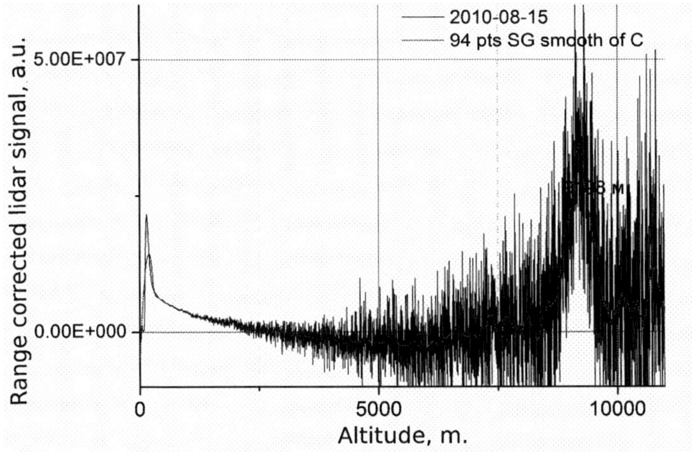

Fig.7 Vertical distribution of light-scattering layers near Matua Island, August 15, 2010

Our research has illustrated that volcanic activity brings significant disturbance into the atmosphere. This disturbance was found as far as 200 km away from the volcano, and was present during at least 2 months after the eruption. Maximum deviations in the atmosphere corresponded to the planetary boundary layer and the tropopause layer where most of the volcanic ash was present. High concentrations of volcanic aerosols in the PBL can stimulate mass intrusions in the top-most ocean layer, which may result in phytoplankton blooming or death. Satellite data analysis revealed that the tropopause layer had a lifetime of about 2 months. The 2010 expedition proved that one year after eruption the atmosphere completely restored its background characteristics.

References

Chen B.B., Sverdlik L.G.. Laser radiation polarization in dust an clouds in Central Asia region // Vestnik KRSU/№ 5, 2003

Khain V.E., Khalimov E.N. Global Climate Changes and Circularity of Volcanic Activity. Science without borders // Transactions of the International Academy of Science H&E. Vol.3. 2007/2008

Zuev V. V., Bukin O. A., Pavlov A. N., Shmirko K. A.. Results of Joint Observations of Aerosol Perturbations of the Stratosphere at the CIS-LiNet Network in 2008 // Atmospheric and Oceanic Optics, 2009, Vol. 22, No. 3, pp. 295–301

Influence of Atmospheric Correction on Calculation Accuracy of Sea Reflecting Capacity in SeaDAS Program Complex

VASILY KACHUR

Observation of the ocean color is of great importance. Phytoplankton being the major contributor to the water color, remote observations of the ocean color allow finding concentration of phytoplankton chlorophyll in water. Nowadays satellite data are the only practically viable way of monitoring spatial and seasonal plankton variations at the water surface level; and they present the most vital information for studying ocean primary productivity, global processes of carbon dioxide consumption and evaluation of the sea ecological condition. This work analyzes accuracy of various atmospheric correction algorithms available in the SeaDAS program complex, and singles out the most appropriate algorithms for processing satellite data of the Gulf of Peter the Great.

Light can go in various ways before it reaches the satellite sensor. Thus, radiance intensity (E) measured by an artificial satellite can be presented as

$$E = E_A + E_M + E_{\text{Î ÒÐ}},$$

where E_A is brightness of the signal dispersed and reflected in the atmosphere; E_M - brightness of the signal coming from the sea depth (the useful signal); $E_{\text{отр}}$ — brightness of the signal reflected by the sea surface. Atmospheric correction of radiance is applied for extracting the useful signal from the entire signal. Atmospheric correction accuracy vitally influences the accuracy of water bioparameters calculated by means of satellite information.

The SeaDAS program complex is widely used for processing satellite data. It is capable of processing data from various sources, including such satellite radioparameters as SeaWiFS, CZCS, MODIS, etc.

The goal of this research lies in comparative analysis of work of various atmospheric correction algorithms available in SeaDAS and singling out the most suitable algorithm for satellite processing of data obtained in the Gulf of Peter the Great.

In order to carry out this analysis, two expeditions (from August 31, 2009 till September 14, 2009 and that of September 29-30, 2009) took in situ measurements of sea upwelling radiation in the Gulf of Peter the Great. These data helped calculate coefficients of sea reflecting capacity (Rrs). On the following days – 31.08.09, 02.09.09, 04.09.09, 08.09.09, 11.09.09, 13.09.09, 14.09.09, 29.09.09 – one hundred sites were selected with established coordinates and their calculated Rrs coefficients (on the broad wavelength range with an interval of 1nm) for further comparison with satellite data. Then, every site within the range was assigned coefficient values with wavelengths on which MODIS satellite radiometer channels we were interested in were focused, i.e. 412 nm, 443 nm, 469 nm, 531 nm, 555 nm, 645 nm, 667 nm, 678 nm and 748 nm. These coefficients were compared with those calculated by the SeaDAS programme complex with use of various atmospheric correction algorithms. It was sea reflecting capacity that was chosen to assess atmospheric correction accuracy, as calculation of the majority of other sea bioparameters includes this one.

SeaDAS has 12 atmospheric models, selection of which for radiance correction is based on the type of waters (shoreline waters, sea waters or ocean waters) as well as relative air humidity data. Our experiment dealt with atmospheric correction algorithms that could choose a necessary model by themselves. Those were: multi scattering NIR correction, multi scattering MUMM correction and single scattering correction.

Satellite data were provided by the MODIS radiometer of Aqua Satellite. SeaDAS allowed calculation of Rrs values in satellite image pixels with the same coordinates as those of in situ measurements (satellite image resolution being equal to 250-1000 meters). These calculations were carried out for each of the three examined atmospheric correction algorithms. As the Rrs calculation algorithm was the same, thus divergence between satellite and in situ data depended on atmospheric correction accuracy. Therefore, comparative analysis of those divergences could reveal which of atmospheric correction algorithms would be more preferable for the Gulf of Peter the Great.

Here and further on:

$$\Delta = R_{rs_{MODIS}} - R_{rs};;$$

$$\delta = \Delta / R_{rs};;$$

Rrs_{MODIS} – reflecting capacity, calculated with SeaDAS, based on MODIS radiometer data;
Rrs – reflecting capacity counted in compliance with in situ measurement results;
Δ – divergence of Rrs;
δ – «relative» inaccuracy of Rrs calculation.

By means of atmospheric correction for all channels of the MODIS radiometer, Δ and δ values for each atmospheric correction algorithm were obtained. These values allowed for plotting correspondent charts. Figures 1, 2 and 3 illustrate examples of such charts. The vertical axis shows δ, the horizontal axis shows sites where measurements were taken. Charts also show the dates when measurements were taken.

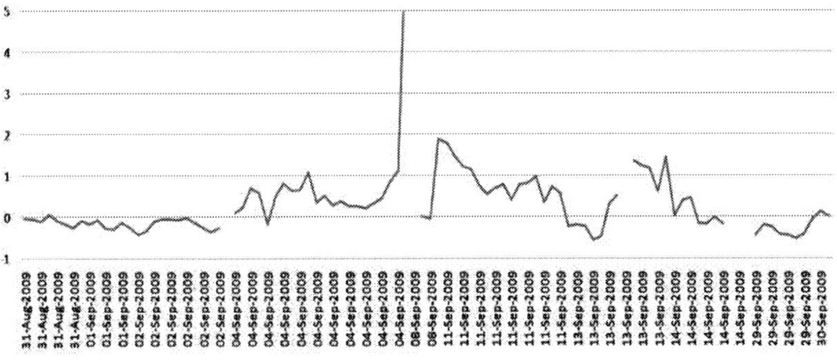

Fig.1. Relative inaccuracy (δ) of Rrs calculation with MUMM correction

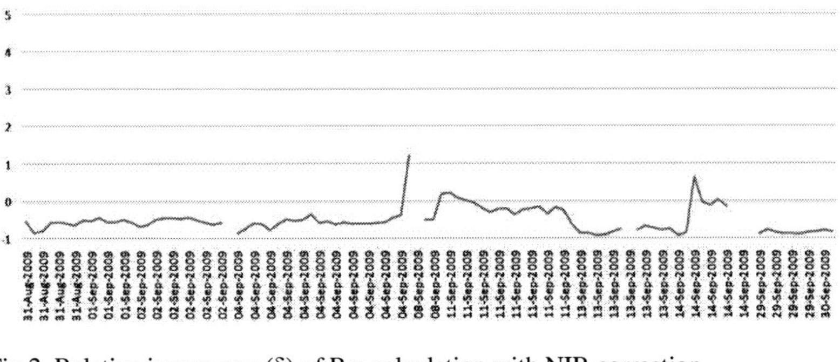

Fig.2. Relative inaccuracy (δ) of Rrs calculation with NIR correction

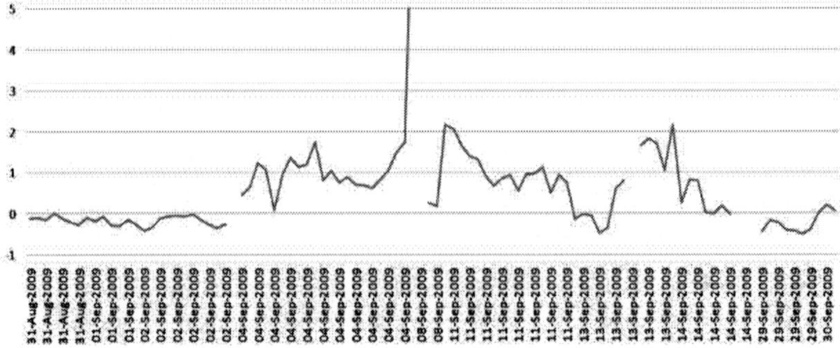

Fig.3. Relative inaccuracy (δ) of Rrs calculation with Single scattering correction

We also analyzed the approximate value, the median and inverse distribution of Rrs calculation inaccuracy in various atmospheric correction algorithms. Figure 4 shows the chart of dependence of the relative inaccuracy spread median on the MODIS radiometer channel.

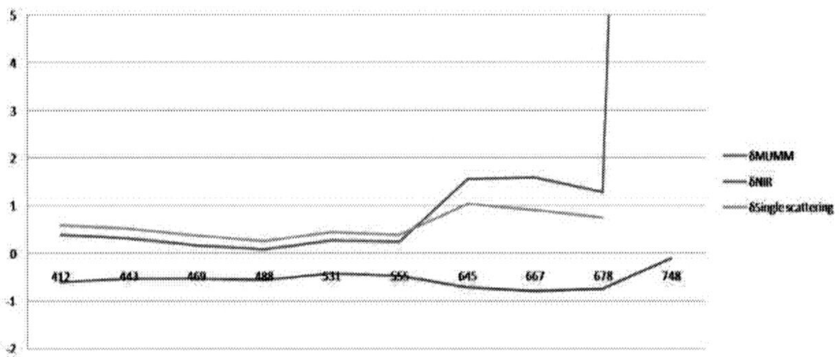

Fig. 4. Median of relative inaccuracy (δ) of Rrs calculation

Analysis of divergence of satellite and in situ Rrs values resulted in singling out a range of regularities, including the below-described ones.

In the majority of cases, MUMM and Single scattering give higher Rrs value than in site measurements. NIR almost always lowers this parameter. Moreover, use of NIR-correction diminishes the volume of calculated Rrs. On channels with the wavelength ranging from 412 nm to 555 nm, MUMM produces the best calculation inaccuracy. When Single scattering correction is implemented, SeaDAS can not calculate Rrs values for the 748 nm channel. Also, it produces slightly smaller inaccuracy than MUMM on channels ranging from 645 nm to 678 nm. On the whole, quite low accuracy of atmospheric correction for the three algorithms on 645 nm, 667 nm, 678 nm, and 748 nm channels. Some points show high inaccuracy values of Rrs calculation. Calculation errors may account for this, or majorly this can be the result of inefficient sorting out the «defective» pixels by SeaDAS.

The tall ship Nadezhda was performing an expedition to the Sea of Japan and the Sea of Okhotsk from August 2 till September 2, 2010. During it we collected complex data on the atmosphere and the ocean. The data obtained will allow estimating both accuracy of atmospheric correction as well as satellite algorithm inaccuracies for major sea bioparameter calculations.

Moreover, we carried out the analysis of dependence of atmospheric correction algorithms on such parameters as the nebulosity albedo, atmospheric optical thickness, Angstrem coefficient, relative air humidity, wind speed, NO2 concentration. Obtained regularities will permit to elaborate on the method of the atmospheric correction algorithm optimal selection.

OCEAN COLOUR REMOTE SENSING

PAVEL SALYUK AND IGOR STEPOCHKIN

Part I. Customization of Semi-Analytical Biooptical Algorithm for Gulf of Peter the Great

Ocean color satellite sounding data and fluorometric passive measurements of the sea brightness coefficient obtained aboard the ship were used in elaboration of a hyperspectral biooptical algorithm for the Gulf of Peter the Great. This algorithm will permit to reconstitute the chlorophyll-A concentration value and the dissolved organic matter absorption coefficient of sea upwelling radiation spectrums obtained aboard the ship. In future, algorithm customization and simplification will allow for its application to the sea color satellite sounding data.

Sea reflecting capacity measurements were carried out with a hyperspectral spectrophotometer aboard the ship, by the procedure based on NASA protocols. Sequentially, we measured non-calibrated brightness spectrums of sea surface upwelling radiation:

$$S_{sfc}(\lambda, \theta, \phi \in \Omega_{FOV}, \theta_0)$$

Sky non-calibrated brightness spectrums:

$$S_{sky}(\lambda, \theta_{sky}, \phi \in \Omega_{FOV}, \theta_0)$$

Non-calibrated radiation brightness spectrums reflected by a horizontal plate (maximally close to the Lambert surface):

$$S_g(\lambda, \theta_g, \phi_g \in \Omega_{FOV}, \theta_0, \phi_0)$$

With a known function of reflected light distribution into two directions:

$$R_g(\lambda, \theta_g, \phi_g \in \Omega_{FOV}, \theta_0, \phi_0)$$

(Fig.1). For further calculations, each spectral series was filtered according to apparent outliers, and after that, the median spectrum and the spectral distribution measurement error were calculated. All calculations were carried out in the MATLAB environment. The sea water reflecting ability spectral distribution

$$R_{rs}(\lambda,\theta,\phi\in\Omega_{FOV},\theta_0)$$

was calculated according to the following formulae:

$$S_w(\lambda,\theta,\phi\in\Omega_{FOV},\theta_0)=S_{sfc}(\lambda,\theta,\phi\in\Omega_{FOV},\theta_0)-\rho S_{sky}(\lambda,\theta_{sky},\phi\in\Omega_{FOV},\theta_0)$$
(1)

$$R_{rs}(\lambda,\theta,\phi\in\Omega_{FOV},\theta_0)=\frac{S_w(\lambda,\theta,\phi\in\Omega_{FOV},\theta_0)R_g(\lambda,\theta_g,\phi_g\in\Omega_{FOV},\theta_0,\phi_0)}{\pi S_g(\lambda,\theta_g,\phi_g\in\Omega_{FOV},\theta_0,\phi_0)}$$

(2)

Where:

$$S_w(\lambda,\theta,\phi\in\Omega_{FOV},\theta_0)$$

is a spectrum of non-calibrated brightness of sea upwelling radiation;

λ – radiation wave length;
θ and ϕ – zenith and azimuth angles at which the spectrophotometer is directed at the water surface;
θ_0 and ϕ_0 – zenith and azimuth angles of the Sun;
θ_{sky} – a zenith angle, linked to θ; θ_g
and ϕ_g zenith and azimuth angles at which the spectrophotometer is directed at a horizontal plate;
Ω_{FOV} – the spectrophotometer lens viewing field.

Figure 2 shows an example of spectral distribution R_{rs}. The signal in the field exceeding 750 nm can be caused by dispersion on air bubbles or suspensions. Its exclusion was carried out according to the following formula:

$$R_{rs}(\lambda) = R_{rs}(\lambda) - median(R_{rs}(750:850))$$
(3)

The signal exceeding 900 nm contains large errors and is out of our consideration. “Plateau” existence within a range from 750 to 850 nm was one of measured spectrum quality criteria.

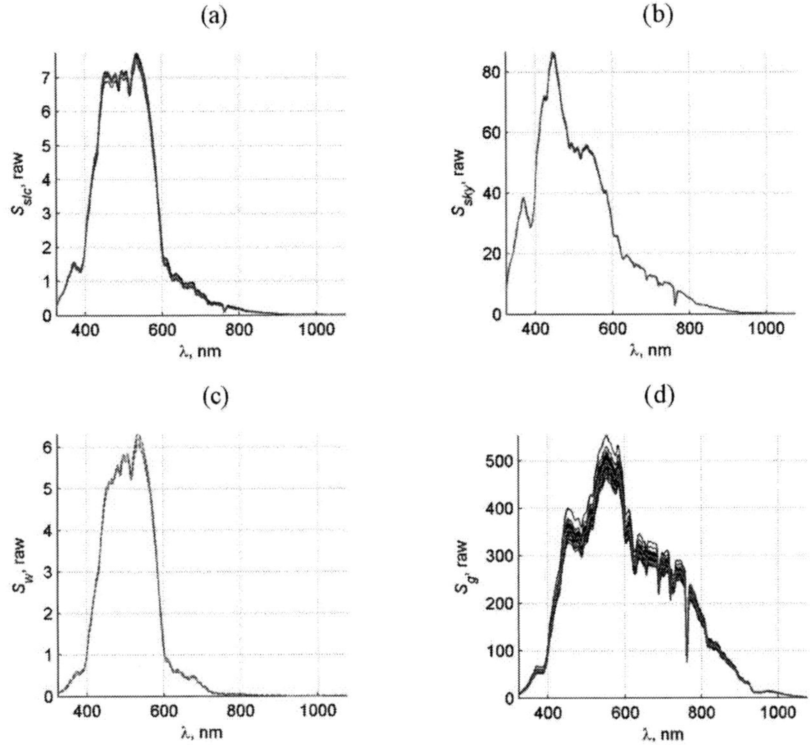

Fig. 1 – Spectrums of non-calibrated radiation brightness: (a) upwelling from sea surface, (b) from sky, (c) upwelling from sea water, (d) reflected from Lambert surface.

The entire definition procedure of one sea reflecting capacity spectrum lasted 10 minutes. To control lighting conditions, PAR (photosynthetic active radiation) was being uninterruptedly recorded (the solar illumination integral signal within the range from 400 to 700 nm). In case of large PAR signal variations, the data obtained were out of further consideration.

Corrective adjustments of the obtained data to the water surface natural viewing angles were carried out in compliance with NASA protocols.

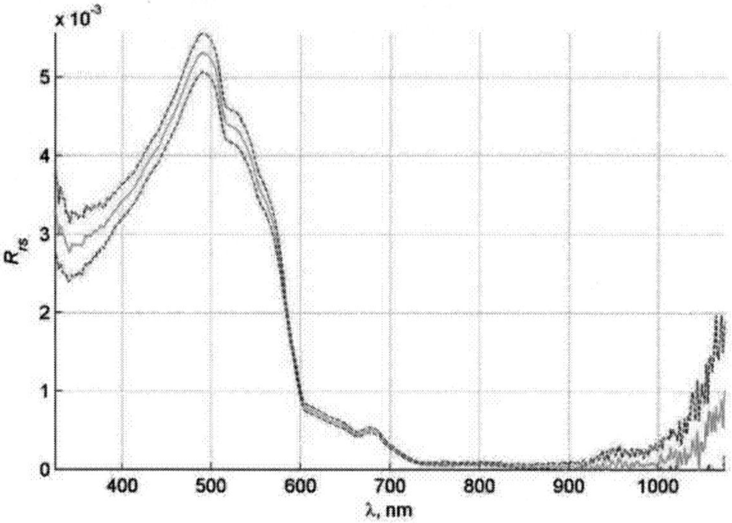

Fig. 2 – Sea surface reflecting ability spectrum.

A portable ASD-manufactured hyperspectral spectrophotometer with an optical path enabling in-water measurements was used for radiometric measurements. Its spectral range is 325 – 1075 nm, its resolution is 1nm. The kit comprises above-water lenses to measure brightness at 10° and 2° viewing angles, an in-water lens to measure brightness at a 5° viewing angle and an in-water lens to measure luminance.

In total, 100 reflective capacity spectrums that complied with our quality criteria were obtained. Figure 3 shows the route of the vessel aboard which the measurements were carried out.

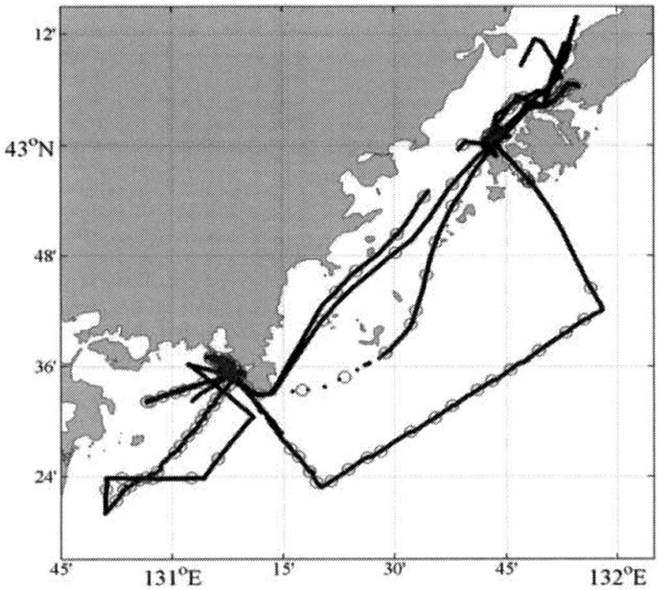

Fig. 3 – Spatial distribution of points where sea water reflecting ability spectrum measurements were carried out.

The following formula was used as a model to show dependence of sea brightness on environmental optical characteristics:

$$R_{rs}^{m} = \frac{\pi \cdot G}{a}\left[\frac{b_{bw}}{3.4} + X\left(\frac{400}{\lambda}\right)^{Y}\right] \tag{4}$$

where X and Y – are spectrum invariables; a – the full coefficient of sea water absorption, its value being described by the sum of model values of pure water absorption coefficients a_w, phytoplankton pigments a_{ph}, organic debris and DOM a_g; b_{bw} – the sea water inverted dispersion coefficient, and G=0.176. The “m” index in R_{rs}^{m} means a model analysis of the sea brightness coefficient.

Sea water absorption characteristics modeling was carried out with a formula for a_g and table values for a_w:

$$a_g(\lambda) = a_g(440) \cdot exp\{-s_g \cdot (\lambda - 440)\} \tag{5}$$

The following model was implemented to describe the absorption coefficient by phytoplankton pigments:

$$a_{ph}(\lambda, C_{Chl}) = A(\lambda) \cdot C_{Chl}^{1-B(\lambda)} \quad (6)$$

where C_{Chl} is chlorophyll-A concentration, and $A(\lambda)$ and $B(\lambda)$ were obtained by means of spline interpolation of tabulated values in the entry. Description of pure sea water scattering properties was carried out within the frame of the model function described by this formula:

$$b_{bw} = 0.0038(\frac{400}{\lambda})^{4.3} \quad (7)$$

Substitution of model functions into (4) permits to find that the sea brightness coefficient is a function of the five parameters: a_g, S, X,Y and C_{Chl}.

Measurement repeatability largely depends on the right choice of initial values and the required parameter search range. Various regression relationships between a_g, S, X, Y, C_{Chl} and $R_{rs}(\lambda)$ were used for their evaluation.

The value obtained from the regional empirical algorithm was used as chlorophyll-A concentration initial meaning (Fig. 4):

$$C_{chl} = 10^{(-1.346R + 0.471)} \quad (8)$$

$$R = \log_{10} \frac{R_{rs}443 > R_{rs}488}{R_{rs}551} \quad (9)$$

The Y value range measurement was evaluated by means of regression relationships given in formula [2]:

$$0.9[0.86+1.2\ln(\chi)] < Y < 1.1[0.86+1.2\ln(\chi)], \quad (10)$$

where $\chi = R_{rs}(440)/ R_{rs}(490)$.

For calculation of the variability interval a_g, the empiric correspondence between chlorophyll-A concentration and a_g was used:

$$\mathrm{Log}[a_g(440)] = -1.20 + 0.47\log(C_{Chl}) \quad (11)$$

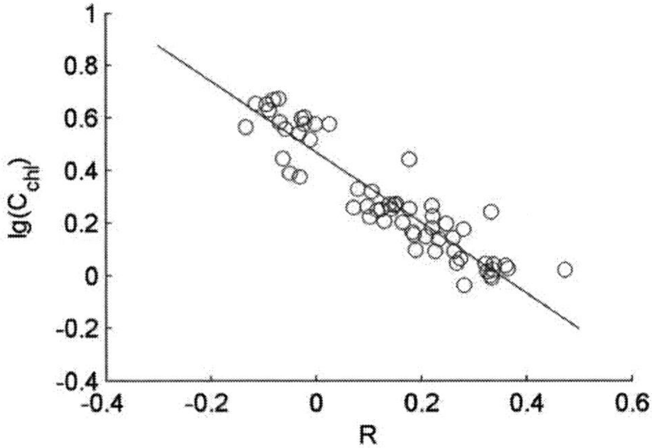

Fig. 4 – Parameter *R* scattering diagram, reconstructed by means of radiometer-obtained data, and C_{chl}, obtained from contact measurements of chlorophyll-A concentration.

Substitution of previously found chlorophyll-A concentration ultimate values into this equation permits to find the variation range a_g. The variable S lies within a narrow range from 0.011 to 0.016, rarely amounting to 0.02. A stipulation that X > 0 is accepted for initial values of the variable X.

Every range of initial values was divided into five points for each unknown parameter. Thus, $5^5 = 3125$ sets of initial conditions were obtained. Further on, the Trust Region optimization algorithm assessed the parameters. Out of 3125 correspondent solutions obtained, only those with the parameter R^2 exceeding 0.99 were selected. Normally, such solutions constituted 60%. Figure 5 shows an example of initial and model spectrums.

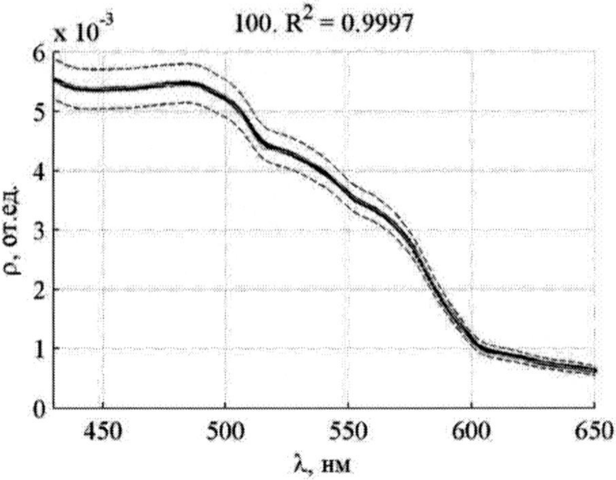

Fig. 5 – Real (black line) and model (gray line) spectrums of sea water reflecting capacity.

In order to separate all "successful" solutions (that have $R^2 > 0.99$) from all "correct" ones (coinciding with ground data), we effectuated the comparison of distributions of chlorophyll-A concentration, reconstituted from reflecting capacity spectrums, and real measurements of chlorophyll-A concentration deep sounding data obtained with the SBE19-plus profilograph. Further on, we chose such ranges of initial conditions that resulted in chlorophyll-A concentration credible values. Figure 6 shows histograms of "correct" initial conditions. These conditions are recommended to be used as limitations for choice of initial conditions. While the initial conditions themselves should be chosen as the most credible value among obtained divisions.

Table 1. Boundaries of "correct" solutions

parameter	1 % quantile	99 % quantile
Cchl	0.8534	5.1629
$a_g(440)$	0.0364	0.5979
S_g	0.0097	0.0175
X	0.0000	0.0073
Y	-6.6616	3.2409

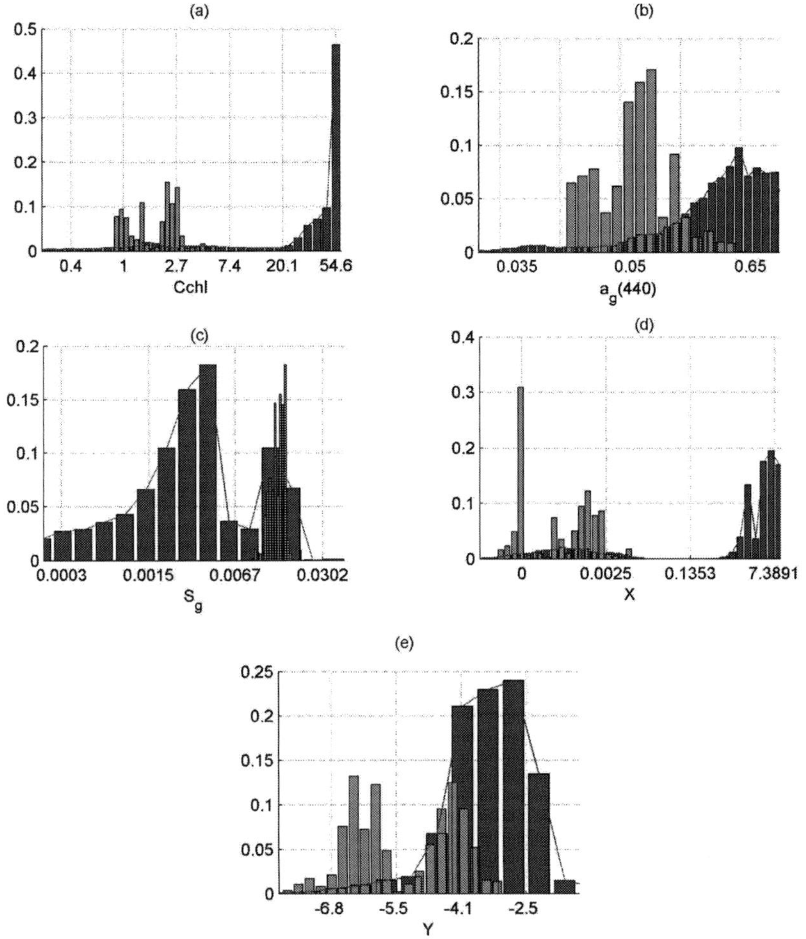

Fig. 6 – Histograms of obtained semi-analytical algorithm solutions. Green color – "correct" solutions, red color – all solutions.

Part II. Dependence of Regional Biooptical Algorithms on Correlation between Chlorophyll-A Concentration and Dissolved Organic Matter

The general approach to the regional empiric algorithm development is confined to choosing one of global satellite algorithms, for instance, OC2, that links chlorophyll-A concentration with a sea brightness spectral components correlation algorithm on wave lengths from 490 to 555 nm; and to searching regression coefficients through composition of satellite and vessel measurements of chlorophyll-A concentration [6]. In elaborating the regional algorithm it should be taken into account that its efficiency will depend on the type of variation of shallow water optical characteristics, and in particular on correlation between proportion of chlorophyll-A and DOM, which can fluctuate both on temporal and spatial scales. Clusters of chlorophyll-DOM scattering diagrams help define algorithm stability areas, and areas of regression coefficients stability respectively. The problem of the level of dependence of regional satellite algorithm stability on characteristics of chlorophyll-DOM scattering diagram clusters remains outstanding, and it is a subject of research of this paper. We are the first to put this approach into practice [7,8].

Figure 7 demonstrates the map of satellite-recorded simultaneous ocean color changes, as well as chlorophyll-A concentration measured aboard a ship with a lotic fluorometer LF3 [9]. The data were obtained within a period from 2008 to 2010 in the Gulf of Peter the Great. All vessel-made measurements were optically measured [10] for correct comparison with satellite data. All vessel measurements were compared with the closest satellite pixel, however, the distance between points could not exceed two km. 1801 synchronous measurements were obtained in total. Satellite data were processed in the FEB RAS Centre for Satellite Monitoring with the help of the SeaDAS programme [11].

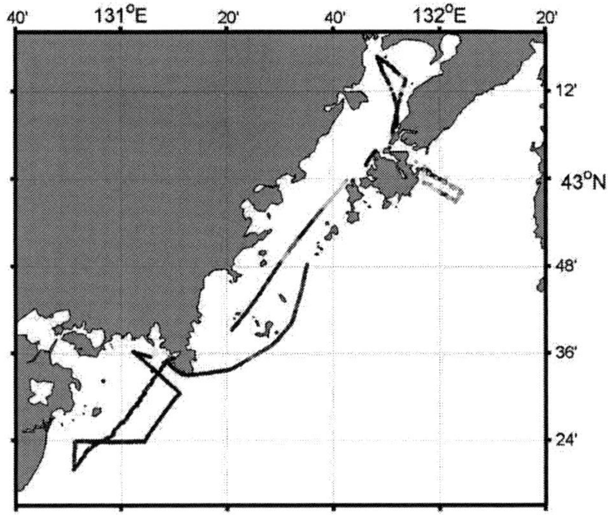

Fig. 7 – Geographic location of satellite and vessel synchronous measurements.

Figure 8(a) shows a diagram of scattering between chlorophyll-A concentration C_{ow} and DOM D_{ow}, measured aboard a ship. Dots to the left from line 1 show areas with stronger DOM influence. Dots to the right side from line 1 show weaker influence. Figure 8(b) presents comparison of water reflecting capacity measured by a satellite and the denary logarithm of chlorophyll-A concentration. Reflecting capacity was counted with the help of formula 9. Line 1 on picture 8(b) is a global algorithm OC3M, line 2 is linear regression traces over all dots (can be used as a regional algorithm). In waters with stronger DOM influence, absorption in the blue spectrum region is also stronger, and the values of R_{oc3m} become excessive in relation to the global algorithm. On the contrary, R_{oc3m} in waters with low inclination value is excessive and dots move to the right in relation to the curved line oc3m on figure 8(b). The shift value is labeled as dR.

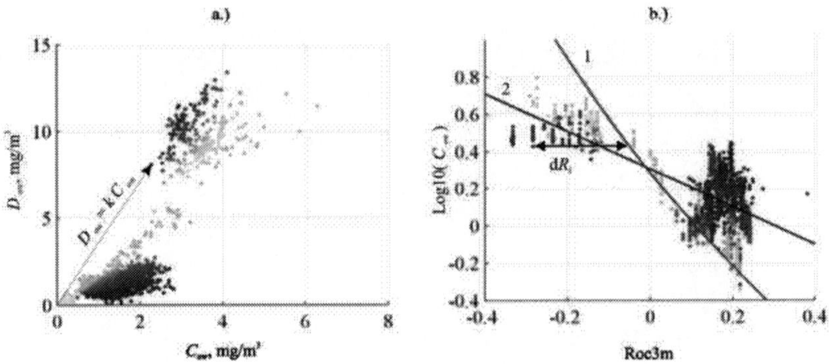

Fig. 8(a) Chlorophyll-A and DOM concentration scattering diagram measured aboard ship; (b) comparison of satellite-measured water reflecting capacity and chlorophyll-A denary logarithm. Value of relation k between chlorophyll-A and DOM concentration is shown in color.

Figure 9 illustrates comparison of the obliquing factor k and the shift dR. Observed is considerable linear dependance, with the correlation coefficient being equal to 0.81.

Therefore, to increase satellite algorithm work precision it is necessary to map the area under investigation according to the obliquing factor and to use the obtained linear correlation of figure 10.

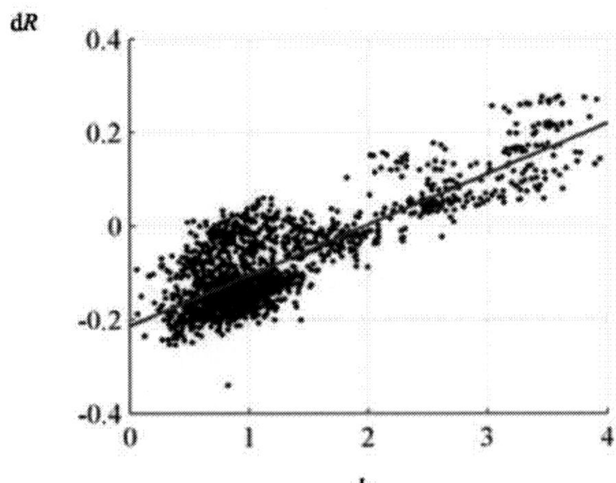

Fig. 10 – Coefficient k and shift dR scattering diagram.

Part III. Functional Check of Elaborated Regional Empiric Biooptical Algorithms Based on Various Bodies of Data

In the course of carried research [13] obtained were the following empirical algorithms that reconstitute chlorophyll-A and dissolved organic matter concentration from satellite-obtained data in using the sea color in the Gulf of Peter the Great:

chl_reg01 – Regional chlorophyll 1

$\log_{10}C_{ChlA} = 1 - a_1 \times \exp(a_2 \times R_{OC3M})$, $a_1 = 0.725$, $a_2 = 2.02$ (12)

$$R_{OC3M} = \log_{10}\frac{R_{rs}443 > R_{rs}488}{R_{rs}551} \quad (13)$$

chl_reg02 - Regional chlorophyll 2

$C^0_{chlA} = \left(\dfrac{a}{\mathrm{FLH}+k} - 0.2\right)^{-1}$ with $R_{OC3M} < -0.1$ (14)

$\log_{10} C_{ChlA} = a_1 + a_2 \times R_{OC3M} + a_3 \times R_{OC3M}{}^2$, a1 = 0.347, a2 = -2.9, a3 = 4, R_{OC3M} >= -0.1 (15)

$$R_{OC3M} = \log_{10}\frac{R_{rs}443 > R_{rs}488}{R_{rs}551} \quad (16)$$

dom_reg01 – Regional DOM 1

$C_{RDOM} = a \cdot C_{OC3M} + b \cdot C_{Rchl2} + c$, a = 0.8646, b = 1.4701, c = -1.1027 (17)

C_{oc3m} = 10^(0.283 - 2.753 *R_{oc3m} + 1.457 * $R_{oc3m}{}^2$ + 0.659 *$R_{oc3m}{}^3$ -1.403 *$R_{oc3m}{}^4$) (18)

Figure 10 shows geographic locations of fluorescence measurements that were simultaneously carried out along with passive optical measurements. Dots designate measurements effectuated simultaneously with satellite sounding, circles – those carried out simultaneously with upwelling sea radiation measurements taken aboard a ship. The "color 1" shows 2008 data, the "color 2" – 2009, the "color 3" – 2010.

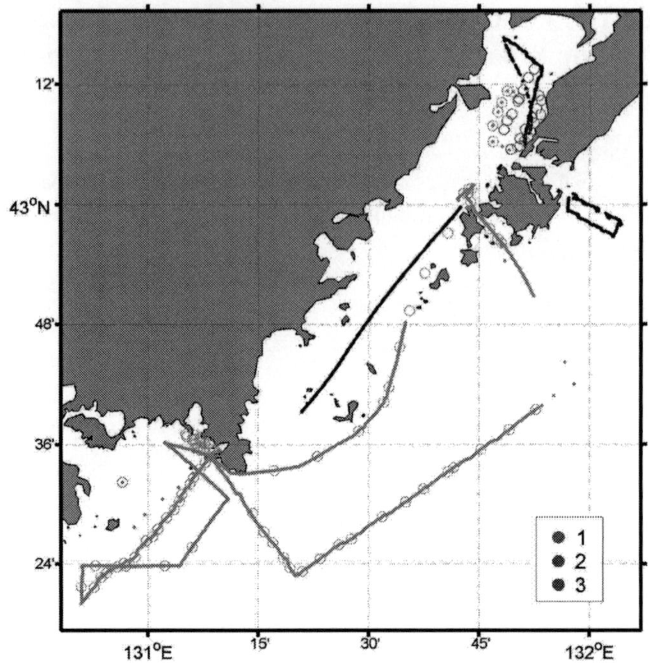

Fig. 10 – Geographical location of simultaneous ship fluorescent measurements and optical passive measurements of upwelling sea radiation in 2008-2010.

Figure 11 shows chlorophyll-A and DOM concentration scattering diagrams, reconstituted from sea color passive sounding data and ship fluorescence measurement data.

Tables 2 and 3 present estimates of usage accuracy of various global and regional biooptical algorithms in relation to various bodies of data obtained over several years. Table 2 shows (in percentage terms) relative values of the y=x dot disalignment mean-square error. This value can be taken for the algorithm accuracy estimate.

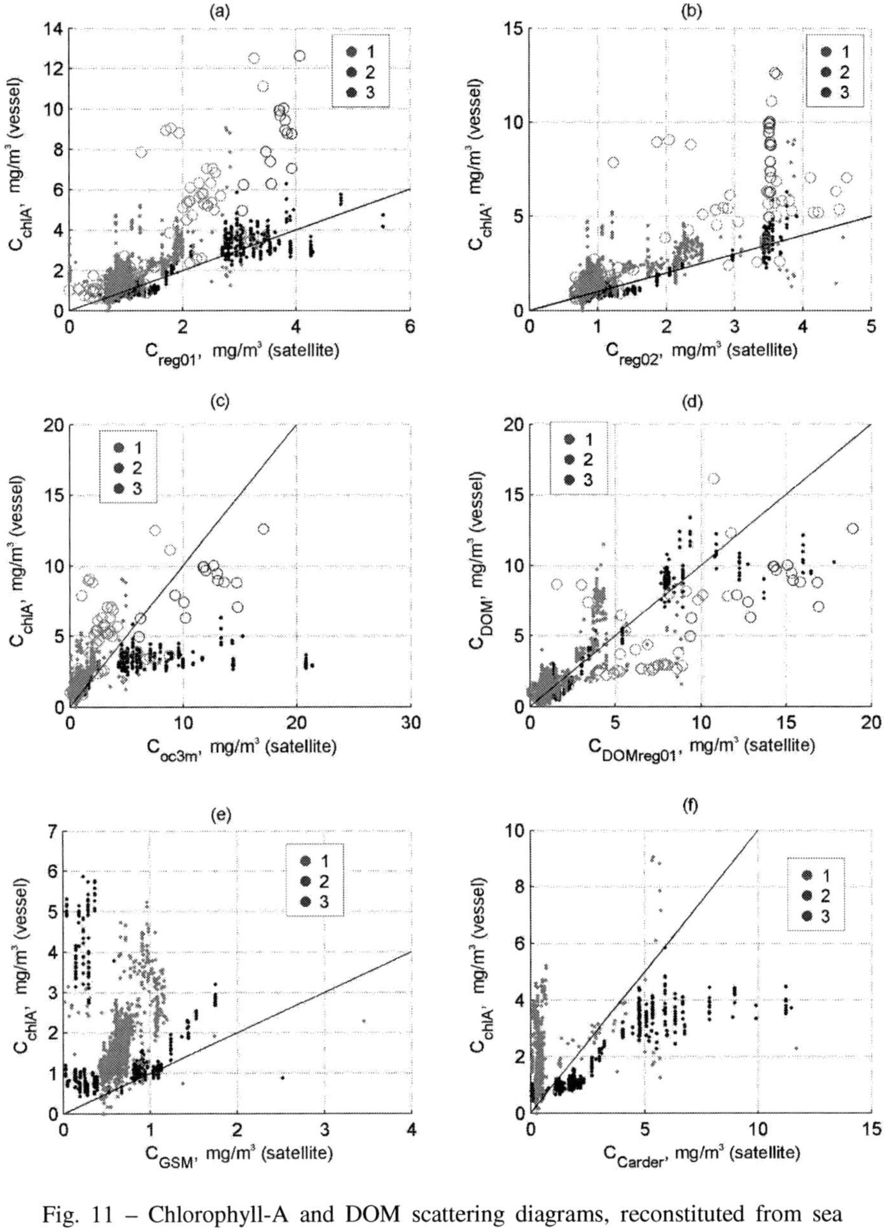

Fig. 11 – Chlorophyll-A and DOM scattering diagrams, reconstituted from sea color passive sounding data and ship-obtained fluorescence data.

Four bodies of data were used for ocean color passive sounding: 2008 Sat MODIS-Aqua satellite sounding data obtained in 2008 (“color 1” dots) and implemented in elaboration of regional biooptical algorithms; 2009 Sat MODIS-Aqua satellite sounding data obtained in 2009 (“color 2” dots); 2009 ASD - satellite sounding data obtained with a spectrometer in 2009 (“color 2” circles); 2010 AS - ship sounding data obtained with an ASD spectrometer in 2010 (“color 3” circles). Table 3 shows values of correlation coefficients between values presented on scattering diagrams, figure 12. Presence of considerable linear correlation testifies of the possibility of carrying out algorithm simple corrections. The “bold” font in tables 2 and 3 shows “good” values of calculated coefficients, the “underline” font – “acceptable” meanings, the “italic” font – “unacceptable” meanings.

The following conclusions can be deduced from analysis of scattering diagrams on figure 12 and tables 2 and 3. Elaborated regional biooptical algorithms show better results than global algorithms when applied to satellite data. Here, global algorithms are oc3m, GSM and Carder. When using ship radiometer data, considerable discrepancies arise. However, algorithms chl_reg01 and oc3m reveal a high correlation coefficient with fluorometer data for algorithms. This testifies of a possibility of making correct regional biooptical algorithms for an ASD ship radiometer. Incorrect performance of chl_reg02 and dom_reg01 algorithms is caused by usage of solar induced fluorescence intensity data, which are non-calibrated in case of usage of a solar radiometer. Thus, values of chl_reg02 and dom_reg01 coefficients need correction if a ship radiometer is used.

The general goal for further research is elaboration of a regional algorithm that will equally properly function for both different year data bodies and various radiometers. Envisaged is a united assessment procedure of solar induced fluorescence line height. Simultaneous use of ship fluorescent measurements and passive optical measurements of sea upwelling radiation will permit to assess atmospheric correction quality.

Table 2. Values of relative errors of analyzed biooptical algorithms.

	2008 Sat	2009 Sat	2009 ASD	2010 ASD
chl_reg01	**27**	73	*113*	84
chl_reg02	**22**	63	*660*	88
oc3m	*200*	87	92	**37**
GSM	*163*	*133*	-	-
carder	84	*137*	-	-
dom_reg01	**40**	*567*	*747*	**52**

Table 3. Values of correlation coefficients when employing various biooptical algorithms.

	2008 Sat	2009 Sat	2009 ASD	2010 ASD
chl_reg01	**0.92**	0.63	**0.75**	**0.76**
chl_reg02	**0.96**	0.65	*0*	0.48
oc3m	0.65	0.61	**0.66**	**0.76**
GSM	*-0.13*	*0.25*	-	-
carder	**0.90**	0.39	-	-
dom_reg01	**0.94**	*0.12*	*0*	**0.76**

References

Baith, K., Lindsay, R., Fu G., Mcclain, C.R., "SeaDAS, a data analysis system for ocean-color satellite sensors", EOS Trans. AGU, 82, 202 (2001)

Bricaud, A., Babin, M., Morel, A. and Claustre, H., "Variability in the chlorophyll-specific absorption coefficients of natural phytoplankton: Analysis and parameterization", J. Geophys. Res. 100, 1321-1332 (1995).

Clark D.K. 2003, Case 1 waters algorithm. MODIS Algorithm Theoretical Basis Document, 18, pp. 1-50

Gordon, H.R., Clark, D.K., "Remote sensing optical properties of a stratified ocean: an improved interpretation", Applied Optics 19, 3428-3430 (1980)

Lee, Z.P., Carder, K.L., Peacock, T.G., Davis, C.O. and Mueller J.I., "Method to derive ocean absorption coefficients from remote-sensing reflectance", Applied Optics 35, 453-462 (1996).

Maior, A.Yu., Bukin, O.A., Krikun, V.A., Baulo, E.N., Lastovskaya, I.A., "Compact flow-type shipborne fluorimetre", Atmospheric and Oceanic Optics 20, 258-259 (2007).

Mueller, J.L., Curtiss, D., Arnone, R., Frouin, R. and Carder K., "Above-water radiance and remote sensing reflectance measurement and analysis protocols", Ocean Optics Protocols For Satellite Ocean Color Sensor Validation 3, 21-30 (2003).

Salyuk, P.A., Bukin, O.A., Pavlov A.N., Mayor, A.Yu., Akmaykin, D.A., Krikun, V.A., "Optical properties of Peter the Great Bay waters compared with satellite", International Journal of Remote Sensing 31(17), (2010)

Salyuk, P.A., Podoprigora, E.L., "Comparative analysis of the chlorophyll A concentrations obtained by the laser-induced fluorescence method (LIF) and SeaWiFS", Proceedings of SPIE, 5149, 53-59 (2003)

Smith, R.C. and Baker, K.S., "Optical properties of the clearest natural waters", Applied Optics 20, 177-184 (1981).

Tassan, S., "Local algorithm using SeaWiFS data for retrieval phytoplankton, pigments, suspended sediment, and yellow substance in coastal waters", Applied Optics 33, 2369-2378 (1994).

COMPARATIVE DATA ANALYSIS OF PASSIVE AND ACTIVE SOUNDING OF BIOOPTICAL PARAMETERS OF SEA WATER TOP LAYER IN SEA OF OKHOTSK AND SEA OF JAPAN IN 2010

VICTORIYA SAKLAKOVA

An ASD upwelling sea radiation spectrometer and a FL3 lotic fluorometer were used for taking measurements during the expedition in 2010. The first apparatus obtains hyperspectral measurements of sea upwelling radiation, which make possible to calculate chlorophyll-A concentration and fluorescence, as well as concentration of dissolved organic matter. The second apparatus measures sea water fluorescence spectrums, which also allows estimating concentration of chlorophyll-A and dissolved organic matter. Comparison of such data will permit to directly develop regional biooptical algorithms without taking atmospheric correction into account. This in turn will allow for optimization of atmospheric correction processes used for ocean color satellite data.

Figure 1 shows the geographic location of analyzed data. White dots show measurements of the upwelling radiation radiometer, grey lines – lotic contact measurements of chlorophyll-A fluorescence.

From the sea water reflecting capacity spectrums obtained with a radiometer (Fig.2), reconstituted were: the parameter R according to the formula

$$R = \log_{10} \frac{R_{rs}443 > R_{rs}488}{R_{rs}551}$$

and the chlorophyll-A fluorescence line height on the 680 nm F_{ASD} wavelength. The parameter R is inversely related to chlorophyll-A absorption, and therefore to its concentration. In a common case, after a molecule absorbs light, photon energy can be used for the following processes: chemical reactions; heating; fluorescence. In its turn, fluorescence depends on chlorophyll-A concentration, solar irradiance, and fluorescence quantum efficiency, which majorly depends on the phytoplankton cell condition. In the case of chlorophyll-A solar-induced fluorescence, relation of concentration of fluorescence must be inversely proportional to solar irradiance and directly proportional to the quantity of photons involved in a photosynthetic reaction.

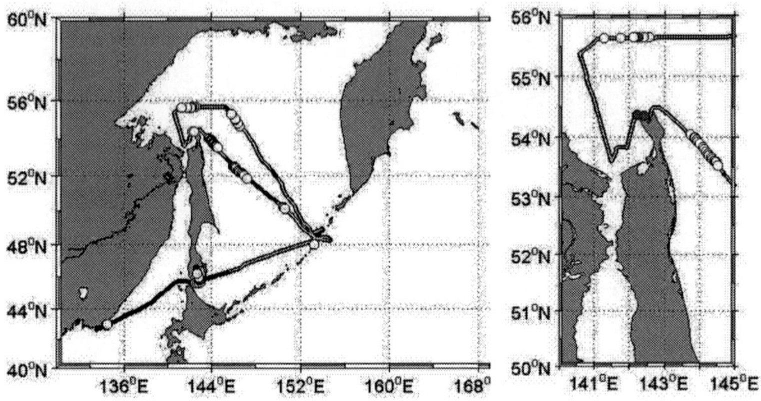

Fig. 1 – Geographic location of measurements carried out with radiometer and fluorometer (white dots and grey lines, respectively).

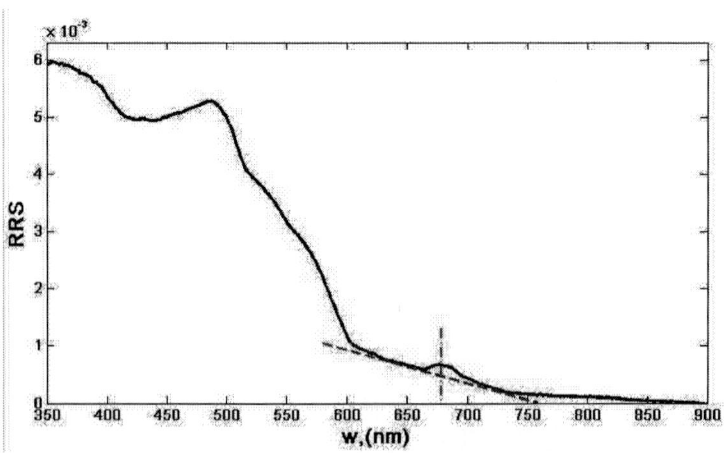

Fig. 2 – Sea surface reflecting capacity spectrum (RRS sr^{-1}).

Figure 3 shows implementation of the regional empirical algorithm, that compares the parameter R (a.u.) with fluorescence intensity, measured with a lotic fluorometer F_{flu}. The following formula describes the points:

$$chl = 10^{-1.54*Roc3m-0.4288}$$

Approximation coefficient equals 0.67.

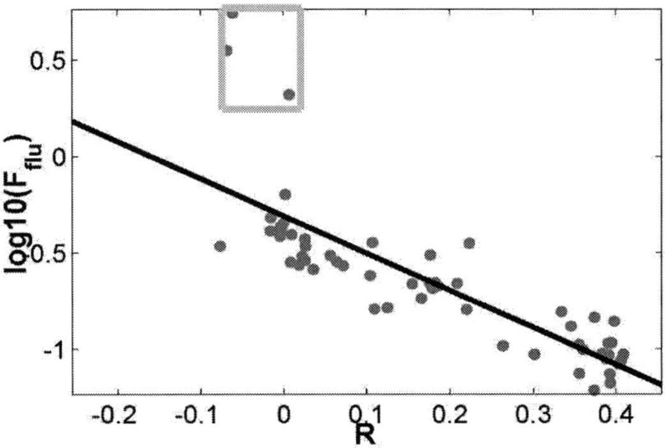

Fig. 3 – Comparison of parameter R, defined through passive measurements of upwelling sea radiation, and contact measurements of fluorescence intensity F_{flu}.

In elaborating the above-described algorithm, the three points that do not satisfy linear approximation were singled out. A rectangle highlights those points. Subsequently, they were out of consideration.

Figure 4a presents a diagram of chlorophyll-A concentration values dispersion, obtained with a radiometer C_{asd}, and chlorophyll-A laser-induced fluorescence F_{flu} values obtained with a fluorometer. The data are well approximated with a direct line, the approximation coefficient being equal to 0.77. Also compared were solar-induced fluorescence values F_{asd}, obtained with a radiometer, and chlorophyll-A laser-induced fluorescence F_{flu}, obtained with a fluorometer (Fig. 4b). Approximation coefficient equaled 0.7.

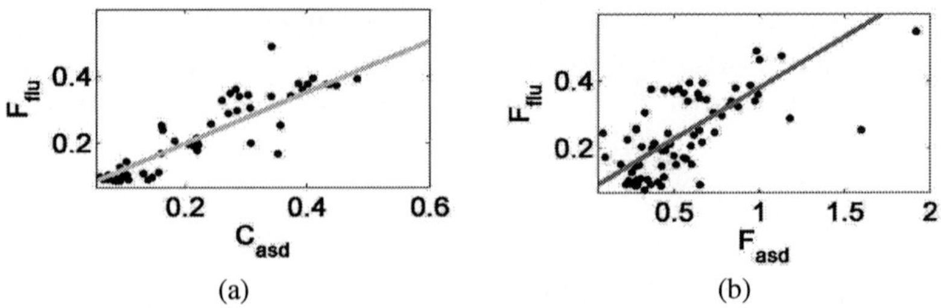

Fig. 4 – Diagram of contact-measured chlorophyll-A concentration values dispersion F_{flu} (a.u.), and (a) remotely measured chlorophyll-A concentration C_{asd} (mg m^-3) reconstituted from sea upwelling radiation spectrums; (b) remotely measured chlorophyll-A solar-induced fluorescence F_{asd} (a.u.).

The given results showcase the possibility of use of remotely measured chlorophyll-A solar irradiant fluorescence data (including satellite-obtained data), in order to increase measurements accuracy or to estimate the phytoplankton state. It is one more way to separate passive remotely sensed ocean color signals caused by dissolved organic matter and chlorophyll-A, because the height of sun-induced fluorescence line very insignificantly depends on dissolved organic matter content in seawater, while ocean color in blue spectral diapason depends on both chlorophyll-a and dissolved organic matter concentration.

References

Abbott M.R., Letelier R.M., Laney S., and Bartlett J.S. Field and laboratory measurements of passive fluorescence and applications to MODIS data // Proceedings of SPIE (Ocean Optics XV) CD-ROM.– Palos Verdes Estates.– CA: SPIE.– 2000.

Fiorani L., Okladnikov I.G., and Palucci A.. First algorithm for chlorophyll-a retrieval from MODIS-Terra imagery of Sun-induced fluorescence in the Southern Ocean // International Journal of Remote Sensing.– 2006.– V. 27.– P. 3615-3622.

Gower J., and King S.Validation of chlorophyll fluorescence derived from MERIS on the west coast of Canada // International Journal of Remote Sensing.– 2007.– V. 28.– P. 625-635.

RECEIPT METHODOLOGY AND PROCESSING OF SEA WATER DISSOLVED ORGANIC MATTER FLUORESCENCE THREE-DIMENSIONAL SPECTRUM

KONSTANTIN KLUGER

The research data obtained on the board of the tall ship Nadezhda within the period from August till September 2010 were used to develop the methodology this article describes. The scientific expedition followed the following route: Vladivostok – Korsakov – Kashevarov Bank – the Amur River estuary.

Spectrafluorimeter optimal setup was optimized to receive a three-dimensional spectrum, as well as the procedure of auto extraction for the sea water three-dimensional spectrum was established, in order to successfully carry out the research activities.

First, sea water samples were measured by the Varian Cary Eclipse fluorimeter. The disadvantage of this device is that it was impossible to measure high intensities received by Rayleigh scattering or its 2nd harmonic. Therefore, a spectral domain with high intensities could not be scanned. Whereas the whole spectrum was divided into the following ranges: 260-510 nm by emanating radiation and 240-300 nm by exciting radiation, 310-610 nm and 290-380 nm, 390-840 nm and 370-500 nm. Emanating and exciting radiation of the monochromator scanning pitch was 5 nm, scanning speed was 3000nm/min and the duration of one measuring was 0,1 sec. Measurement data were saved in an inconvenient way for further usage. To solve this problem MATLAB was used and a particular algorithm was applied. To receive data to be thoroughly analyzed the procedure combined three spectral ranges into one spectrum by overlap averaging 10 nm wide.

Having caddied out all these operations, the fluorescence three-dimensional spectrum was obtained (Figure 1). Wave lengths of emanation and excitation correspond to the abscissa and ordinate axes respectively, while colour gradation illustrates fluorescence intensity.

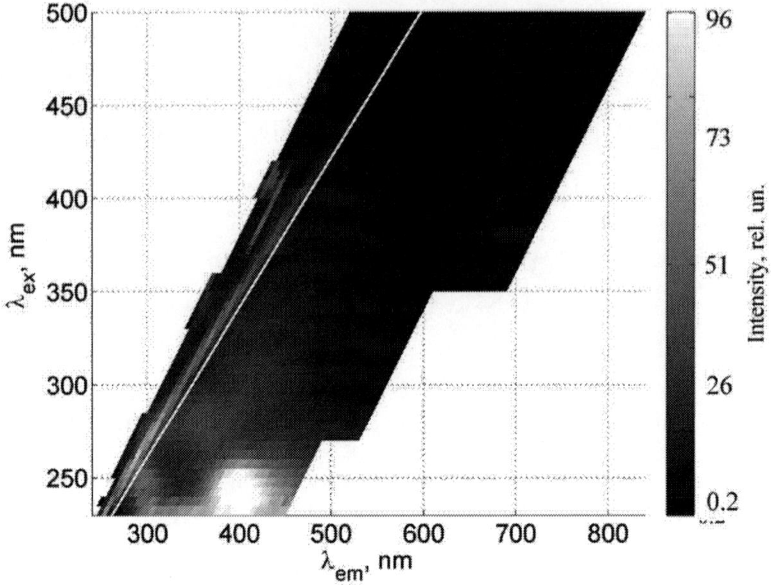

Figure 1 fluorescence three-dimensional spectrum

Considering its complex structure, data processing was carried out as follows (Figure 2):

1. The water Raman scattering (RS) line was cut off. A direct line was traced at the RS line (Figure 1), so the spectral domain above the direct line was not analyzed. It was done in order to take away fluorescence maximums that are close to the Ramen scattering line. The abovementioned maximums are considered to be instrumental, therefore everything above the direct line was not analyzed;
2. Two-dimensional weighted smoothing of the whole spectral region was calculated by Gaussian function. This step was applied in order to reduce noise influence that can cause additional maximums in the time of processing;
3. Weak signals filtration. In order to give a better review of maximums we are interested in, the search was made by intensity of each point. Then, all points with more then level 2 for each spectrum were chosen from the whole lot of points. The remaining points were not analyzed;

4. Detection of maximums on the spectrum edges. Whereas it was impossible to scan the spectrum lower 230 nm by exciting radiation, data arrays of intensity and exciting and emanating wave length were enlarged. This operation made possible to determine two-dimension maximums on the spectrum edges.

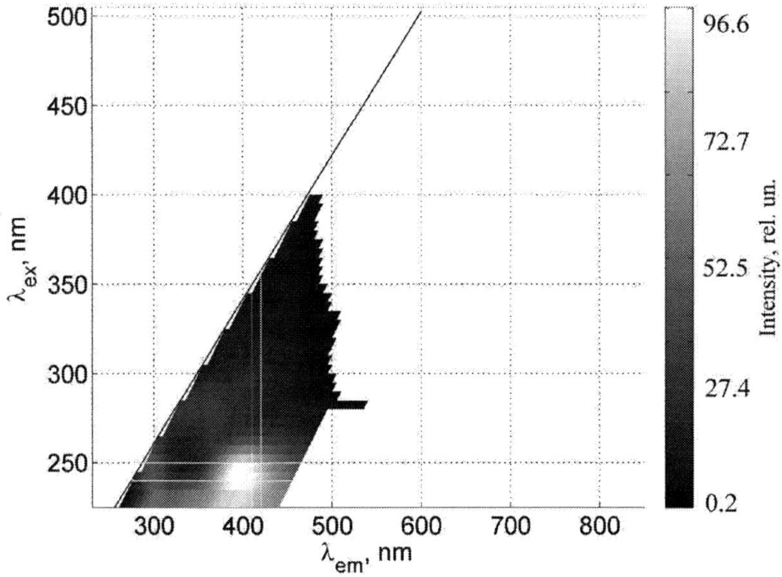

Figure 2 – Spectrum after all steps of processing

The Figure above shows two spectrum regions, which were horizontal and vertical running windows, and it was possible to extract exciting and emanating spectrums. These regions correspond to two-dimension graphics by the length of exciting and emanating waves respectively at a constant horizontal and vertical component of the wave length. Then, three-point maximums were extracted automatically and were designated as peaks (Figure 3). We used the abovementioned algorithm instead of the five-point one or more, in order not to lose maximums on the edges, or in cases when peaks are situated very close to each other.

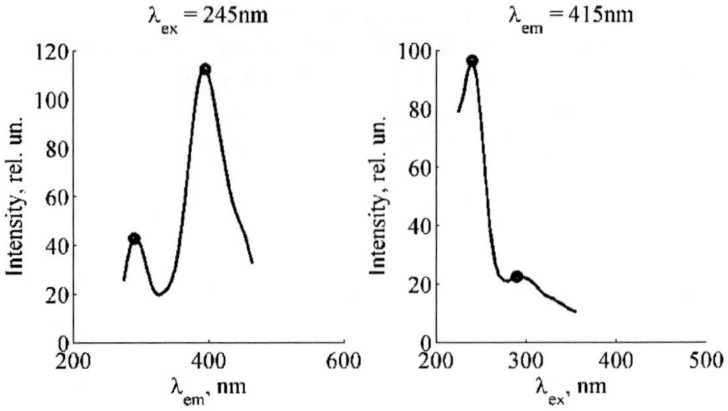

Figure 3 Two-dimension Figures extracted from three-dimensional spectrum

Coordinates of these peaks were recorded on separate arrays, which were drawn later on single axes. On Figure 4, asterisks are for exciting radiation peak maximums, crosses – for emanating radiation. Intersection points coloured in black show peak maximum placement in the three-dimensional space.

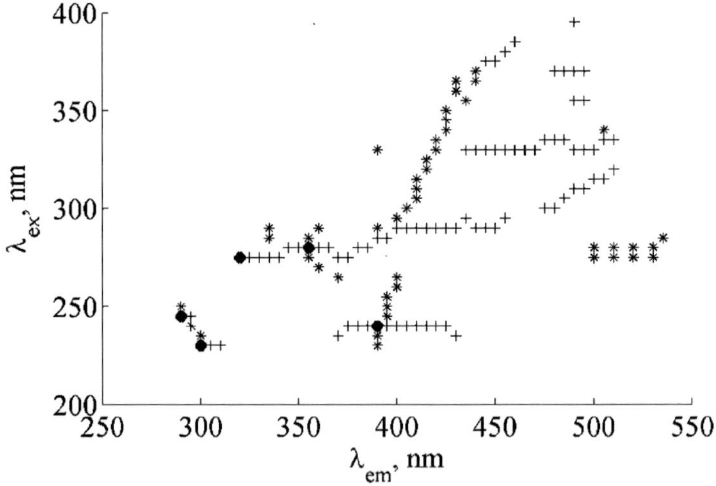

Figure 4 - Example for maximums determination of single three-dimensional spectrum fluorescence

For the reason of a big quantity of samples, a special algorythm was developed in order to show determined three-dimensional maximums on the same graphic – black points on Figure 4. Figure 5 illustrates emanation wave length on the abcissa axis and exitation wave length on the ordinate axis. Colour change (black points) shows a number of hits in this grid region. Grid pitch is 5 nm horisontaly and vertically. Rectangles were used to show the placement and separate out regions of maximums appearance. Figures on the chart show a number of coincedences that came to the dedicated region.

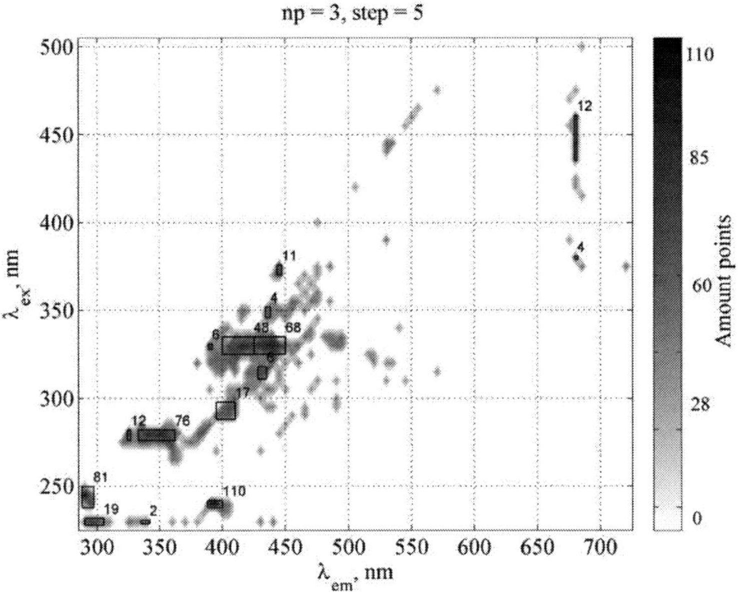

Figure 5. Two-dimension histogram of the three-dimension maximums placement

All results received on the board of the tall ship Nadezhda were compared with research that we had carried out in the past. For example, 2009 sea water measurements were taken in the Bay of Amur coastal waters and in the Golden Horn Bay. Figure 6 presents compared charts. Rectangles show peaks extracted on the analized graphics during the scientific the Nadezhda expedition. It should be noted that some peaks placements are the same, while other peaks have great differences.

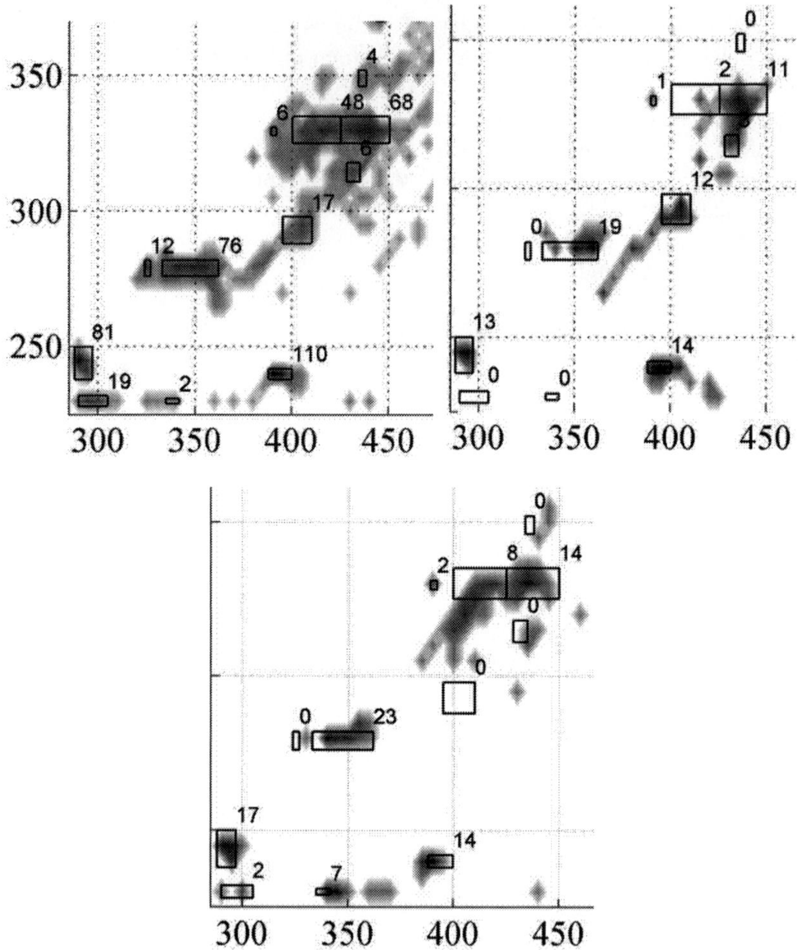

Figure 6 Two-dimension histograms of three-dimension maximums. From left to right: tall ship Nadezhda expedition (Amur Bay and Golden Horn Bay).

All peaks were separated by means of automatical processing for sea water fluorescence three-dimensional spectrums; the abovementioned peaks are shown on the two-dimensional histogram (Figure 5). We succeeded in identifying some peaks from the whole number of extracted maximums (Table 1). The other determined zones with the maximum number of points placement which are absent in Table 1, are currently being identified, for the reason that none of these peaks are beyond description.

Table 1: Identified determined peaks

Class/name of the peak	Excitation wave length, nm	Emanation wave length, nm
Protein similar tyrosine [B peak]	229-231	290-305
Protein similar tryptophan [T peak]	230	340
Humic similar Visible-terrestrial [C peak]	320-360	420-460
Humic similar Visible-marine [M peak]	288-298	395-410
Chlorophyll A [P peak]	435-460	680

SPATIOTEMPORAL DISTRIBUTION OF DISSOLVED ORGANIC MATTER OF VARIOUS TYPES IN SEA OF OKHOTSK

IVAN DOROSHENKOV

Research of dissolved organic matter (DOM) and processes of its transformation in the ocean is essential from the viewpoint of assessment of phytoplankton cell photosynthesizing apparatus work efficiency, interaction between major climate-shaping factors and phytoplankton communities, and assessment of carbon cycle parameters in the upper ocean. First of all, an opportunity of using phytoplankton communities for influencing the dynamics of greenhouse gases in the lower atmosphere depends on how upper ocean DOM fields will be reacting to the dramatic increase of chlorophyll-A concentration and how they will achieve the dynamic equilibrium in the carbon cycle.

DOM is a complex heterogeneous combination of aliphatic and aromatic polymers with the particle size not exceeding 0.2 μm. Originally, it was called "yellow matter" or "humic substance". Later on, it was defined as dissolved organic matter (DOM) and this definition appears to be more precise. Though both definitions – DOM and humic substance – belong to the same chemical reaction of vegetal and algal decomposition, sources of their origin can differ. Thus, for instance, the use of the term "DOM" is restricted to the matter obtained in natural water, while humic substance can enter water environment from outside [1,2]. The definition of DOM emphasizes more precisely that not only visible light is absorbed, but also ultraviolet emission, such as UV-A (λ=315–400 nm) and UV-B (λ=280–315 nm), which is a part of the radiated solar spectrum [3,4]. Therefore, DOM is an essential constituent of the water system that controls penetration depth of harmful solar radiation into the water layer. Thus, the quantity of DOM on the water surface can have a strong impact on the level of harmful radiation that aquatic organisms receive. This can be of vital importance in areas with thinning stratospheric ozone layer in Polar Regions. Also, DOM is of vital ecological importance, being nutritious matter for phytoplankton, micro zooplankton, and bacteria [5]. Thus, it contributes to the gas balance of the entire planet. Phytoplankton communities are estimated to be one of the chief sources of oxygen production on the planet, accounting for some 50% of all oxygen produced and absorbing as much CO2. Thus, even a small disruption of this production process or a long-term quantity decrease of ocean DOM can result in a considerable misbalance between oceanic and atmospheric CO2. All these aspects raise much interest in oceanic DOM research.

The goal of this research is to single out various kinds of dissolved organic matter that influence biooptical parameters of the Sea of Okhotsk. The spatiotemporal analysis of dissolved organic matter was carried out, its measurements being obtained within a period from August to September 2010 in the Sea of Okhotsk during the work of the Far Eastern Sailing University. The expedition embraced extremely various offshore zones. The following varieties of water structure can be singled out:

- Pacific type of sub-Arctic structure – Pacific waters, carried by the Kurile Current;
- Sea of Okhotsk type – water of the Sea of Okhotsk, characterized by especially low minimal temperatures in the cold middle layer and an underdeveloped warm middle layer;
- Southern part of the Sea of Okhotsk type – water of the Sea of Okhotsk, distinguished by high parameters of thermohaline characteristics in the upper level resulting from penetration of the Soya Current into the southern part of the Sea of Okhotsk;
- The Kurile Islands zone type (the Kurile variety) – transformed waters, characterized by different thermohaline parameters in the upper layer (lower temperature parameters and higher salinity ones in relation to neighboring sea and ocean waters), more vertically-strong cold middle layer and smoother extremums of water mass;
- Shallow water zone – waters possessing almost homogeneous vertical distribution of thermohaline parameters.

This water division can affect DOM concentration and composition in the given areas. The expedition carried out from August 2 to September 2, 2010, has revealed different behavior of fluorophores – fluorescent dissolved organic matter. During this period, DOM fluorescence intensity in more than 70 seawater samples was researched on the following route: the city of Korsakov – the Krusenstern Strait – the Kashevarov Bank – the Amur River estuary (Fig. 1). Also, a pump laser hyperspectral fluorometer was constantly recording DOM and chlorophyll-A fluorescence (excitation radiation wave length – 355 nm, spectral range from 365 to 700 nm). The SBE-45 thermosalinograph provided temperature and salinity data. Intensity of various fluorescers was assessed by three fluorescence spectrums (Fig. 2), obtained with the Varian Cary Eclipse electron-tube multifrequent spectrofluorometer (working spectral range of excitation radiation is 240-500 nm, emitted radiation is 250–750 nm, scanning monochromator pace – 5nm). Figure 2 shows main types of DOM, where H is fluorescence of humin-like substances, P - fluorescence of protein-like substances, NP1- unidentified peak.

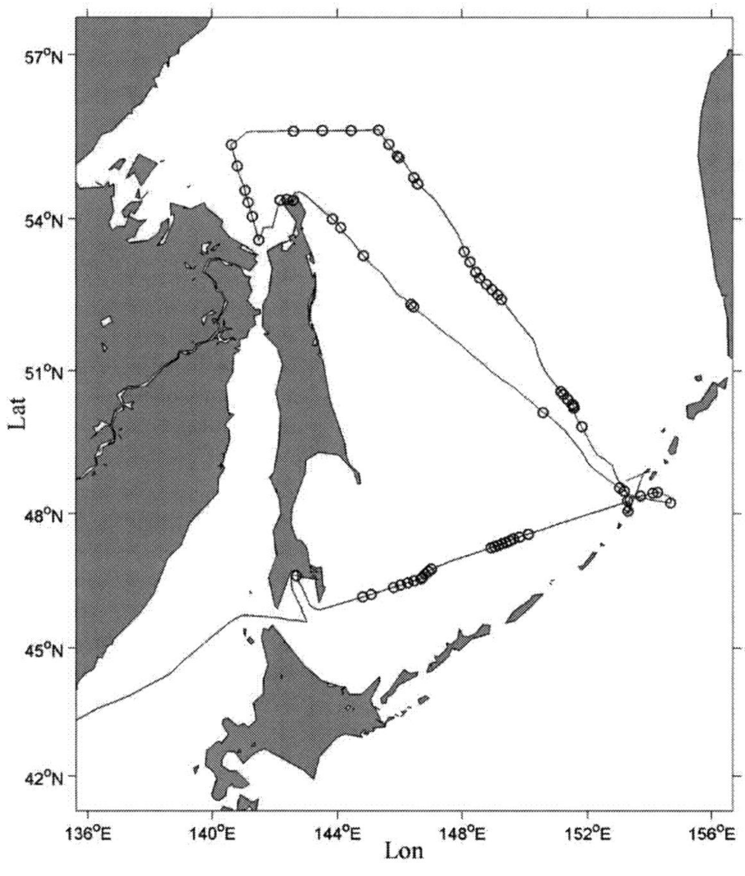

Fig. 1 – Expedition route and sea water sampling points for three-dimensional fluorescence spectrum measurements.

Analysis of various dissolved organic matter types indicated that correlation coefficients of humin-like matter areas of H class constituted 0,97-0,99, which shows similar origin of these fluorescents. That is why further study of humin-like fluorescents can be carried out at one of the peaks.

Reliability of obtained data is confirmed by correlation coefficients of humin-like peak H1 (λ_{em} = 230 - 260, λ_{ex} = 370 - 440) and DOM fluorescence measured with a lotic laser fluorometer (λ_{em} = 355 nm) which equals 0.81. Further analysis used only those points that were described by the obtained linear regression.

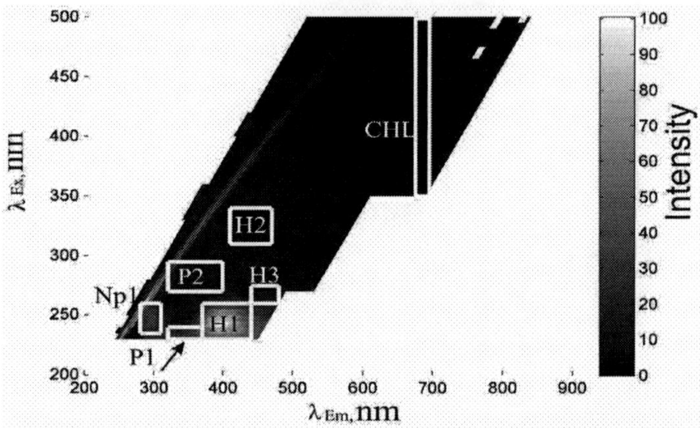

Figure 2 – Three-dimensional fluorescence specter obtained with three-dimensional Varian Cary Eclipse spectrofluorometer. H1,H2,H3 represent class of humin-like fluorescers; P1,P2 – class of protein-like fluorescers; CHL – chlorophyll-A peak; NP1 – unidentified peak.

A scattering diagram of NP1 and H1 peaks (Fig. 3) reveals 2 areas that led to definition of major areas of the Sea of Okhotsk, according to their DOM composition (Fig. 5). Area 1 includes the northern part of the Tartar Strait, the Amur River estuary, and the Kashevarov Bank area. Area 2 – the southern part of the Sea of Okhotsk and the Kruzenstern Strait.

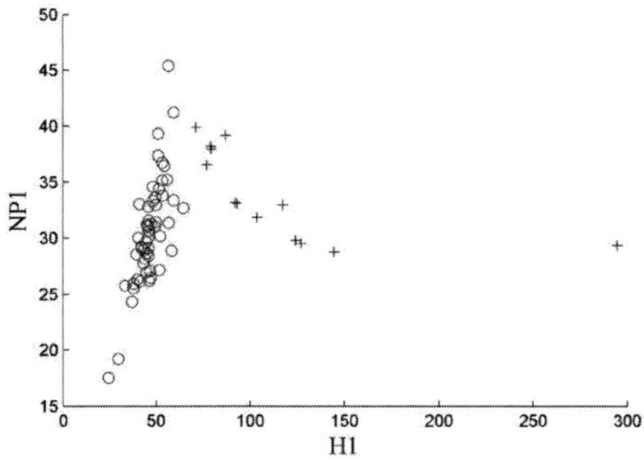

Fig. 3 – Diagram of H2 and NP1 scattering, crosses show Area 1, circles show Area 2.

Area 1, strongly influenced by sea currents and the Amur River, has several subareas with various DOM fluorescence intensity behaviors. Fluorescence of humin-like H1 has high inverse correlation with salinity ($R^2 = 0,86$) (Fig. 4) which testifies of the high inflow of humin-like matter with the Amur River fresh water current. Chlorophyll-A peak (CHL) has direct dependence on H1 ($R^2 = 0,71$), which can be attributed to either inflow of freshwater phytoplankton species, or to the fact that humin-like matter is a chlorophyll-A waste product (Fig. 5).

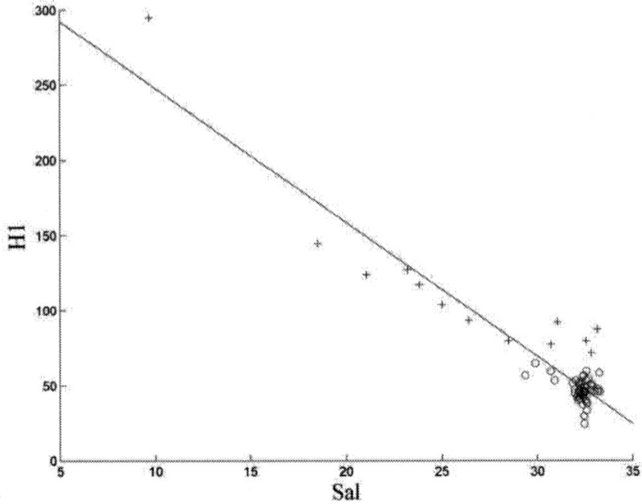

Fig. 4 – Dependence of humin-like H1 on salinity.

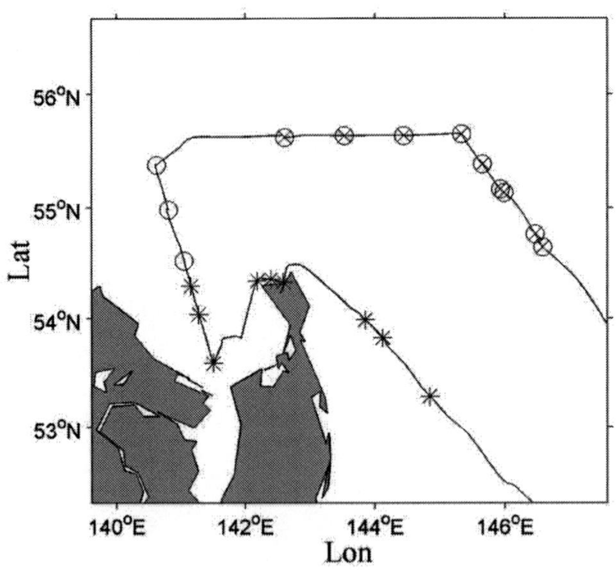

Fig. 5 – Route map of Area 1: a) ⊗ - places influenced by Amur River and pure waters of Sea of Okhotsk b) O- waters slightly influenced by Amur River c) ✳ - waters strongly influenced by Amur River.

Area 2 analysis did not reveal any particularities of fluorescent behavior. Therefore, in the course of the works, the two major areas with various behavior of dissolved organic matter were singled out. The northern part of the Sea of Okhotsk with a considerable amount of phytoplankton is proved to be influenced by different water types: the pure waters of the Sea of Okhotsk and those of the Amur River.

References

Azam F.T., Field J.G. Gray J.S. Meyer–Reil L.A., Thingstad F. The ecological role of water-column microbes in the sea // Mar. Ecol. Prog. 1983. Ser 10, pp.257 -263.

De Mora S., Demers S., Vernet M., The Effects of UV radiation in the Marine Environment // Cambridge University Press, Cambridge 2000.

Neale P.J., Kieber D.J. Assessing biological and chemical effects of UV in the marine environment: Spectral weighting functions. In “Causes and Environmental Implications of Increased UV–B Radiation” (R. E. Hester, R. M. Harrison.) // Issues in Environmental Science and Technology № 14, pp. 61-83. The Royal Society of Chemistry, Cambridge, UK. 2000.

Thomas-Smith. T., Blough N.V., Photoproduction of hydrated electron from constituents of natural waters // Environ. Sci. Tecnol., 2001. 35, pp. 2721 – 2726.

Williams P.J. Incorporation of microheterotrophic processes into the classical paradigm of the planktonic food // web. Kiel Meeresforsch. 1981 5, 1 – 28.

FEATURES OF SOUND SCATTERING ALONG NON-UNIFORM TRACES IN THE SEA OF JAPAN AND IN THE SEA OF OKHOTSK

ANDREY STOROZHENKO

This article presents the results of sound scattering research along nonuniform traces in the Sea of Japan in the Sea of Okhotsk. The purpose is to determine the hydrophisical characteristics of small-scale inhomogeneties of sea medium and their spatial structure in the sea upper layer. Measurements of the sound scattering coefficients in the upper layer of the sea had been performed by the high frequency narrow beam sonar, which was maintained on the nose of the vessel. The complicated spatial structure of the small-scale inhomogeneities of the upper layer of the sea is revealed and characteristic variations of thermocline related with frontal zone and internal waves of different scale are analyzed. Estimations of biomass distribution on a base of sound scattering were obtained along different traces in the Sea of Japan in the sea of Okhotsk. The obtained results open new possibilities for the application of acoustic methods based on sound backscattering for monitoring of variability of small-scale structure in areas of the World Ocean with strong inhomogeneities like the frontal zone, currents and intensive fishery.

Research of sound scattering characteristics is of interest in order to assess the structure of the sea medium. This possibility is especially important near the frontal zones and borders of currents, when any contact methods do allow for detailed spatial observation of medium structure near the borders. Application of remote methods for sound scattering research in the top layer of sea water in a shelf zone and the shallow seas is of importance. Numerous researches of sound scattering along traces in the Sea of Japan were regularly carried out on research vessels the RV Impulse and the RV Malachite in 1998-2008 in its shelf zone with an exit on shelf border in the Bay of Peter the Great. In August 2010, during the expedition on the tall training ship Nadezda the sound scattering research in the Sea of Okhotsk was carried out. Traces along which sound scattering was investigated, were characterized by the presence of frontal zones, borders of currents and complexity of relief. Thus, sound scattering characteristics of the sea medium have changed significantly on borders of these inhomogeneities and therefore allowed visualizing the structure of such inhomogeneities with a good spatial resolution.

Experimental methods and Instrumentation

The method of sound scattering coefficient measurement in subsurface sea layer was based on narrow beam sound radiation, reception of backscattered sound signals, analog input, recording and preprocessing of the acoustic information by computers. A part of acoustic measurements has been executed with the Acoustic Doppler Current Profiler ADP application (made by SONTEC, USA) which allowed to also examine sound scattering at the frequency 250 KHz. The ADP uses three beams oriented 25^0.off the vertical axis, equally spaced at 120^0. relative azimuth angles. Pressure of radiation was $P \sim 10^5$ Pa/re 1 m, the beam width was θ ~ 2^0, pulse duration τ=1.4 ms. Moreover, an additional acoustic sound scattering measurement system for other frequencies in the range from 100 kHz to 200 kHz was used. In particular, some measurements aboard the tall training ship Nadezhda were carried out at the frequency of 100 kHz. It included a channel of sound radiation with various frequencies, piezoelectric transducers, a receiving chain, a system of input, and a preprocessing of the acoustic information. The navigation system on the basis of GPSMAP-130 provided a fitting of scattering data on traces to geographical coordinates. This method has been tested in many expeditions and this article reflects its main features.

Basic formulas for sound scattering coefficient m_V

For the description of sound scattering in the medium with micro-inhomogeneities, the concept of scattering coefficient m_V was introduced, Born approximation, which is defined according to the following formula:

$$I_{bs} = I_i V m_V / r^2, \tag{1}$$

where $I_i \sim P_i^2$, $I_{bs} \sim P_{bs}{}^2$, are the intensity of incident sound radiation on inhomogeneity and that of scattered sound accordingly, thus P_i and P_{bs} are amplitudes of incident wave on volume V and the backscattered wave, accordingly, r is a distance to inhomogeneity, V is a pulse scattering volume of medium. For the high-directivity emitters operating in a pulse mode the volume equals $V = \pi r^2 \theta^2 c\tau / 2$, where θ is the beam width, c is sound speed, τ is a sound pulse duration. Formula one makes it possible to obtain the experimental value of sound backscattering coefficient in a liquid in the form of

$$m_V = (2/\pi\theta^2 c\tau)\left(P_{bs}/P_i\right)^2 . \qquad (2)$$

Using the obtained equations it is possible to evaluate experimentally the frequency dependence $m_V(\omega)$, which allows to identify the type of scatterers, and also some of their characteristics, for example, distribution function in the sizes $g(R)$ or the volume concentration V_0 defined as $V_0 = (4\pi/3)\int_{R_{min}}^{R_{max}} R^3 g(R)dR$.

Formula 1 and formula 2 have been taken as a principle of definition of sound scattering coefficient on the basis of experimental data by measurements of amplitudes of the incident wave on volume V and scattered in the opposite direction, P_i and P_{bs}, accordingly, and formula 2 is a basis for an estimation of the biomass distributed in sea water. The general expression which allows m_V to obtain biomass data within a frame of two distributions, the power and the Gaussian distributions, on the base of experimental data is:

$$m_g^{(P,G)}(r,f) = D^{(P,G)}(f)m_v(r,f), \qquad (3)$$

where the factor $D^{(P,G)}(f)$ has a complicated form. For typical frequency f~250 KHz, it is possible to obtain $D^{(P)}$ ~$2\cdot10^3$ g·m/l from. The resulted estimation is used for an evaluation of biomass along traces.

Basic experimental results

Figure 1 presents the record of sound scattering layers at the frequency of 250 KHz along the trace from Cape Schultz to the deep sea at the end of the summer – the beginning of the autumn 2007. On the right below, there is a map of the route from Cape Schultz towards the deep sea (the length is 20 km). Variability of structure of the subsurface layer, characteristic depths of the thermocline along the trace, as well as sound scattering caused by internal waves are clearly visible.

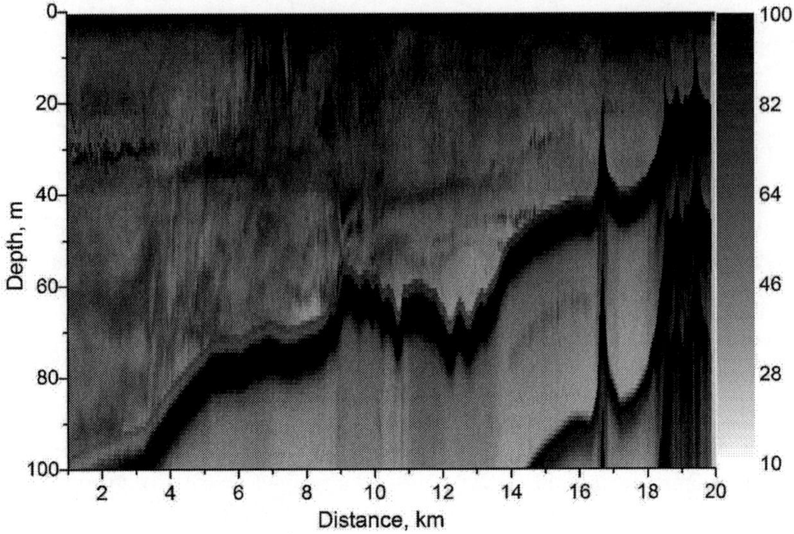

Fig. 1. Sound scattering at frequency 250 kHz on trace of about 20 km (end of summer – beginning of autumn 2007).

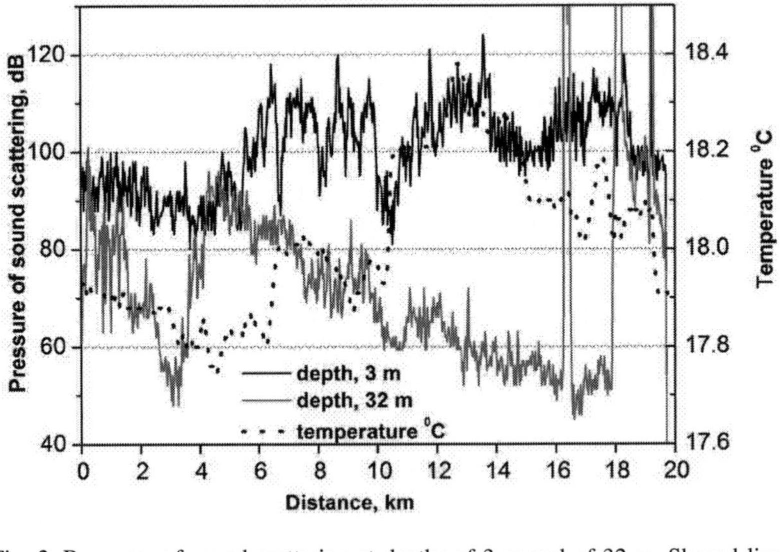

Fig. 2. Pressures of sound scattering at depths of 3 m and of 32 m. Shaped line: temperature distribution at the depth of 2 m along all trace (Fig. 1).

Figure 3 presents the record of sound scattering layers at the frequency of 100 kHz along the trace 52 km in the Sea of Okhotsk (near Cape Elizabeth, northern end of Sakhalin Island). The differences in character of sound scattering are visible from comparison of the results that Figure 1 and Figure 2 present. Sound scattering layers in the Sea of Okhotsk are stronger in comparison with the shelf of the Sea of Japan. On Figure 3 the variations of sound scattering are visible, which is caused by internal waves located near upper and lower borders of thermocline on depths 4-6 m and 10-15 m.

Possessing sound scattering data, it is possible to use their values for estimation of distribution of plankton biomass, which presumably brings the basic contribution to the sound scattering. For this purpose, it is necessary to use formula 2 and 3, which allow calculating distribution $m_g(r,z)$ for various depths along the entire trace. Here we will give an integrated estimation of the biomass along the trace, calculated on depth according to the formula $m_g(r) = \int_0^{z_{max}} m_g(r,z)dz / z_{max}$, where z_{max} is the maximum depth. Figure 3 presents the results. It is visible that there is a considerable spatial variability $m_g(r)$, which is connected with specific plankton depth distribution along the trace presented on Figure 3. Modulation of the thermocline by these processes leads to essential variations $m_g(r)$.

Conclusion

Complicated spatial structure of the small-scale inhomogeneities of the upper layer of the sea is revealed and characteristic variations of thermocline related to the internal waves are analyzed in the work. Estimations of biomass distribution on a base of sound scattering were obtained along different traces in the Sea of Japan in the Sea of Okhotsk. The obtained results open new possibilities for the application of acoustic methods based on sound backscattering for monitoring of variability of small-scale structure in areas of the World Ocean with strong inhomogeneities like frontal zone, currents and intensive fishery.

This research was supported by the Russian Foundation for Basic Research (grant No 09-02-01030-a) and the Far Eastern Branch of Russian Academy of Sciences (grants No 09-I-П17-04 and No 10-III-B-07-189).

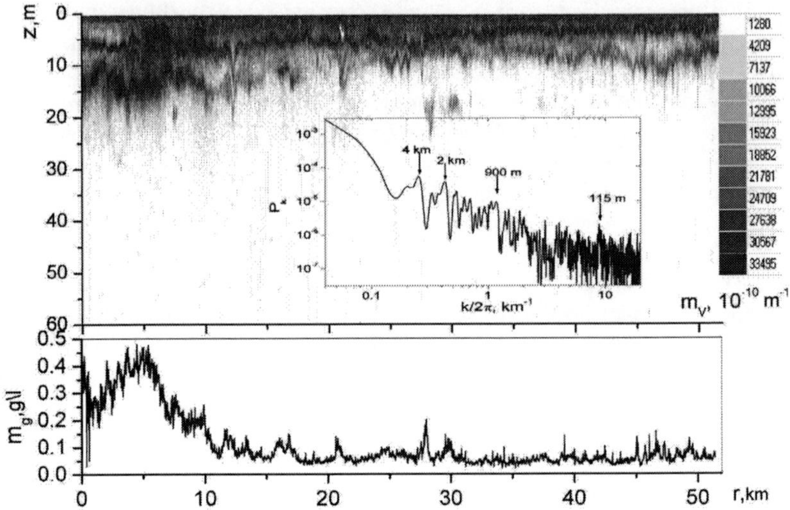

Fig. 3. Sound scattering at the frequency of 100 kHz along trace of the 52 km in the Sea of Okhotsk (near Cape Elizabeth, northern end of Sakhalin island) in August 2010. Lower figure: spatial variability $m_g(r)$ (g/l) along the trace in the Sea of Okhotsk.

References

Andreeva I.B.. Sound scattering layers as acoustic inhomogeneities in the Ocean. Acoustical Physics. 1999. V. 45. №4. p. 385-393.

Bulanov V.A., Korskov I.V., Popov P.N., Storozhenko A.V. "Acoustic sounding of structure of small-scale inhomogeneties in the shelf zone of the sea of Japan"// In: Proceedings of the XXII Session of the Russian Acoustical Society. V. 2. Moscow: GEOS, 2010. p. 265-269 (in Russian)

Bulanov V.A. "On the estimation of biomass distribution in an active layer of ocean using the data of sound scattering"//Underwater investigation and a robotics, №1 (5). 2008. p.58-65 (in Russian)

CORRELATION RELATIONSHIP BETWEEN CHLOROPHYLL-A CONCENTRATION AND ATMOSPHERIC AEROSOL ABOVE KASHEVAROV BANK AND KRUSENSTERN BANK

ALEXEY BOBRIKOV

Not only does the objective of climate-forming factors research imply the direct study of their dynamics and spatiotemporal distribution, but also the study of their interaction with other radiation-active components. Atmospheric aerosol and phytoplankton are among these components. Aerosol direct influence on the condition of photosynthesizing systems of the Far Eastern seas was studied in the work of [Bukin, Shmirko], who indicate that after aerosol emission into the ocean upper layer, there occurs a dramatic rise in chlorophyll-A concentration. Meanwhile, research of phytoplankton contribution into aerosol field formation processes presents considerable difficulties, as the way this influence is carried out is still not entirely clear. There is a hypothesis stating that in the course of their life activity phytoplankton cells produce dimethylsulphide which acts as condensation nuclei and must lead to formation of near-surface aerosol (mists, oceanic haze).

One of the goals of the expedition around the Sea of Okhotsk and the Sea of Japan, carried out from August 3 till September 2, was an attempt of proving or disproving the presence of an inverse relationship between aerosol fields and phytoplankton communities via alternative methods. Thus, regular measurements of biooptical parameters of sea water and optical atmospheric parameters (boundary layer height, average extinction coefficient for the atmospheric boundary layer) were taken aboard the ship.

Kashevarov Bank and Krusenstern Bank dramatically stand out among all the data obtained. It is there that abrupt changes of chlorophyll-A concentration, planetary boundary level height, and extinction coefficient were observed.

Kashevarov Bank and Krusenstern Bank are underwater banks with their tops being 100-150 meters below the water surface. Despite their different geographic position and climatic conditions, areas of thick fog, and chlorophyll-A high concentration on boundaries were observed on both banks. This makes Kashevarov Bank and Krusenstern Bank a suitable area for testing a possible link between phytoplankton and atmospheric aerosol.

As mist formation above the banks can be caused by both meteorological processes and processes related to life activities of phytoplankton communities, analyzed were data obtained with a ship echo sounder, a fluorometer, a temperature sensor set at the depth of 4m and the height of 50m, as well as lidar data. A ship monofrequent aerosol lidar was used for regular atmospheric sounding in long series with accumulation of 100 and 1000 profiles, depending on the atmospheic conditions. Sounding was not carried out when thick lower layer clouds of fog were present in the atmosphere.

The following methods and algorithms were used for dealing with the tasks set. Calculation of the boundary layer height was carried out via lidar signal S-function analysis.

$$S(z) = P(z)z^2 \qquad (1)$$

Here, P(z) is intensity of the received lidar signal without its background component; z is the distance the received signal traveled. Within the limit of the first three kilometers, in the sliding window of 11 dots an area with maximal dispersion was being looked for, then the process was repeated for the found 11 dots with the sliding window comprising 3 dots. The found maximum was defined as the planetary boundary level (PBL) height. This method is more stable compared with the classic method of PBL definition based on the derivative.

Average value of extinction coefficient was found by means of the Beer-Lambert Law. Here, we have made several assumptions.

Extinction (α) and backward scattering (β) coefficients have constant values within PBL. This statement is true, as the atmosphere within PBL limits are quite mixed.

We neglected effects of multiple scattering.

As a result, the following formula is used to calculate the mid-coefficient of extinction:

$$\alpha = -\frac{1}{2\Delta r}\ln\left(\frac{I(r)}{I_0(r_0)}\right) \quad (2)$$

Where $I_0(r_0)$ is the intensity value per 1 dot higher than the backward scattering maximum signal; α is the extinction coefficient ; $\Delta r = r - r_0$ is a vertical layer extension, its length being calculated as α.

Kashevarov Bank is located on the northern shelf of the Sea of Okhotsk. Particularities of local currents and bottom configuration that favor the upwelling of water masses from the depth result in very low water temperature in this area during the summer. Figure 1b shows dynamics of water temperature and air temperature. Moreover, upwelling water brings phytoplankton to the surface, which results in formation of chlorophyll-A peaks on the bank boundaries. Figure 1a shows chlorophyll-A concentration dynamics and bottom configuration.

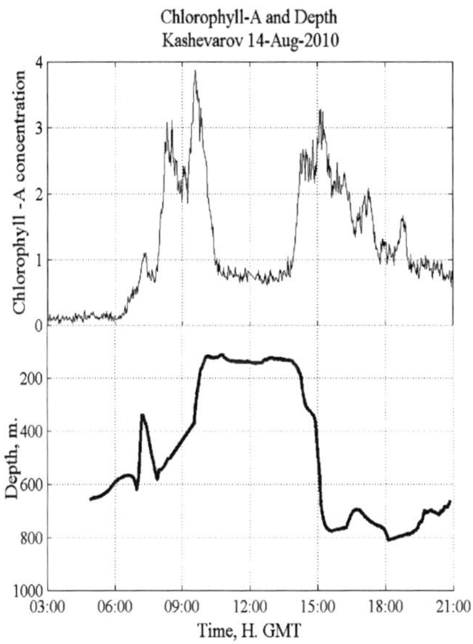

Fig. 1a Chart of chlorophyll-A temporal concentration (top) and bottom configuration chart (bottom).

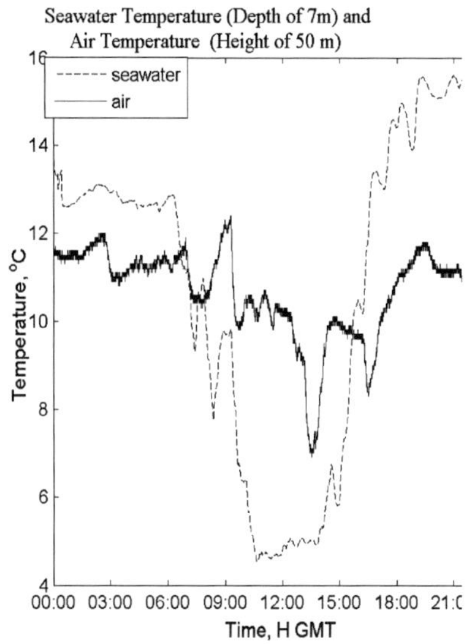

Fig. 1b Water (dashed line) and air (heavy line) temporal temperature changes.

Figure 2 shows PBL height for a chosen area respectively.

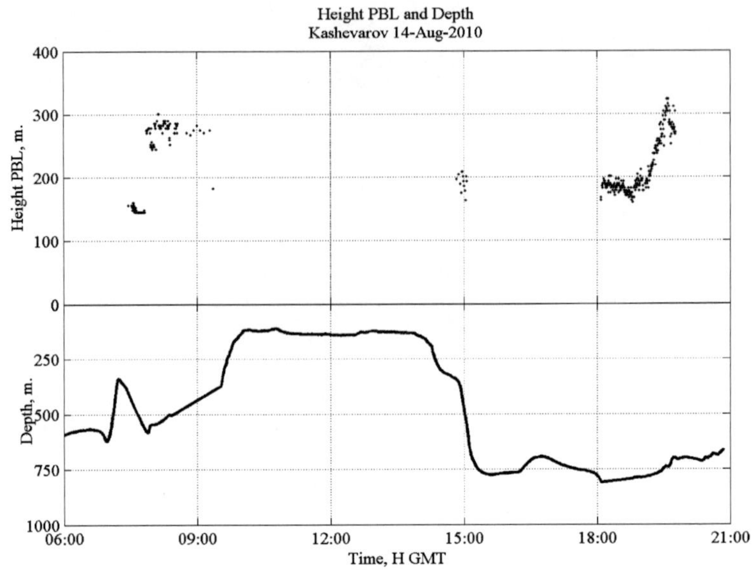

Fig. 2 Temporal changes (dots) of PBL height and depth (line) above Kashevarov Bank, August 14, 2010.

Measurements taken at about 3 p.m. are of insignificant numerical strength, and those obtained after 6 p.m. were located too far from the bank, so both were out of our further consideration.

To achieve our goal, we considered correlation ties between extinction coefficient which is directly tied to aerosol concentration, and chlorophyll-A concentration, water temperature, air temperature, heat fluxes and humidity.

Figures 3a and 3b show scattering diagrams that clearly indicate two relations that correspond to different ocean and atmosphere conditions.

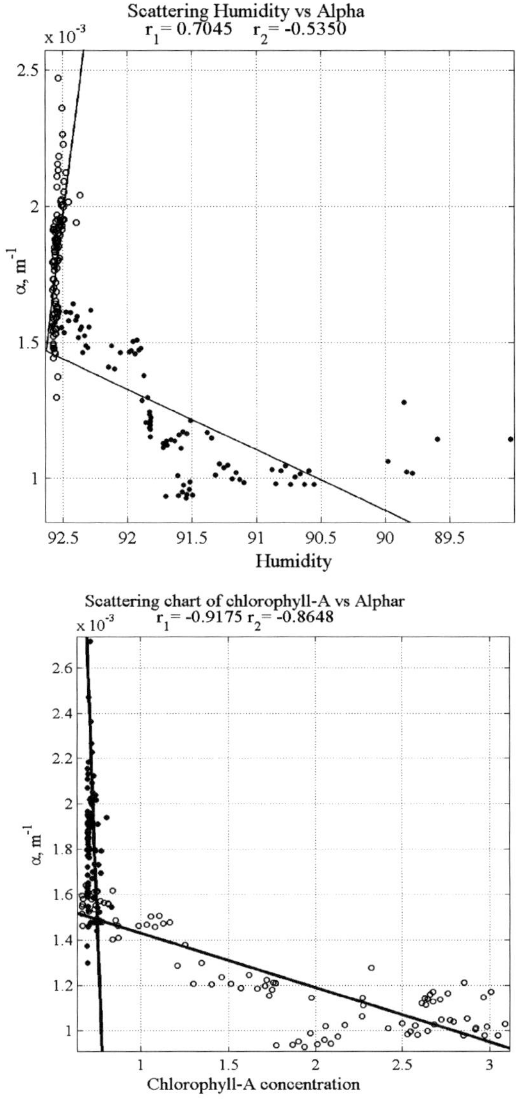

Fig. 3a (top) Scattering diagram of chlorophyll-A concentration and extinction coefficient and Fig.3b (bottom). Scattering diagram of relative humidity and extinction coefficient

The first curved line corresponds to the situation when the extinction coefficient was abruptly dropping down, while chlorophyll-A concentration was slightly going up. This was happening from 7:30 p.m. till 7:40 p.m. (GMT). Then, the speed of extinction coefficient decreased and went down and was steady from 7:40 p.m. till 9:15 p.m. (GMT). Correlation coefficients were: -0.92 for the first time span, and -0.864 for the second time span. This correlation can speak in favor of the following hypothesis: dimethylsulphide, acting as concentration nuclei, leads to growth and condensation of moisture drops in the atmosphere, and thus the atmosphere becomes clearer. On the other hand, the presence of good correlation (0.70 and -0.54 correlation coefficients respectively) does not contradict this theory. Neither does it prove it, however. Insignificant drop-downs in relative atmospheric humidity can be the result of the water drop condensation process.

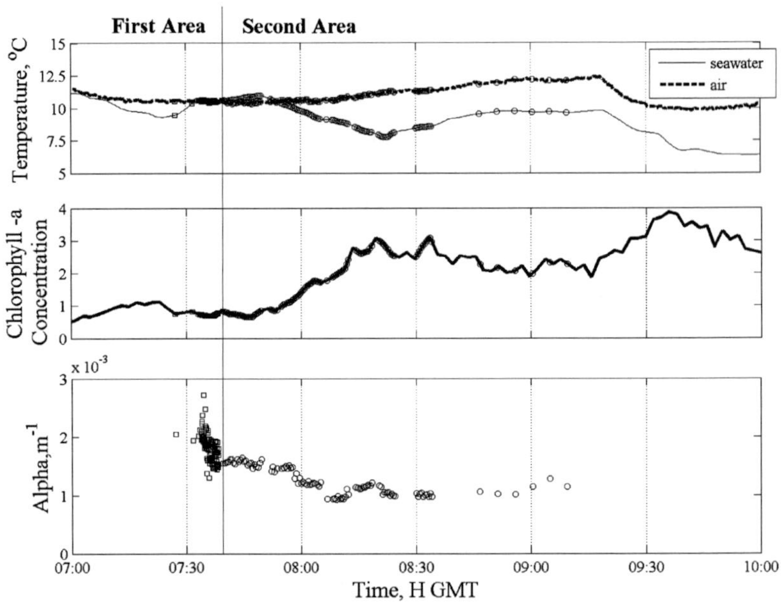

Fig. 3b Data clustering.

Krusenstern Bank is located in the ocean to the east of the Krusenstern Strait. Figures 4a and 4b show its dynamics of chlorophyll-A concentration and water temperature.

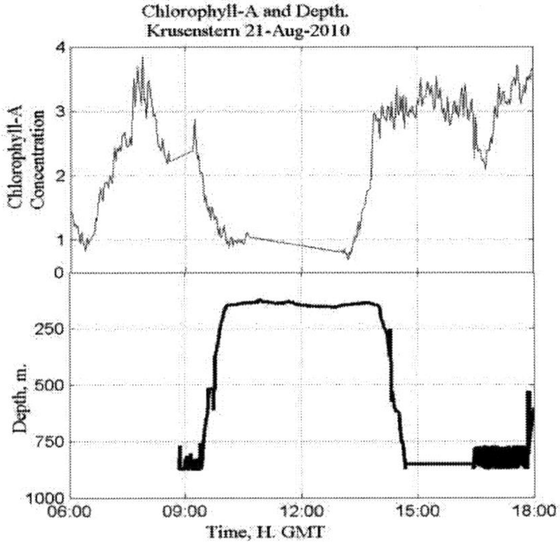

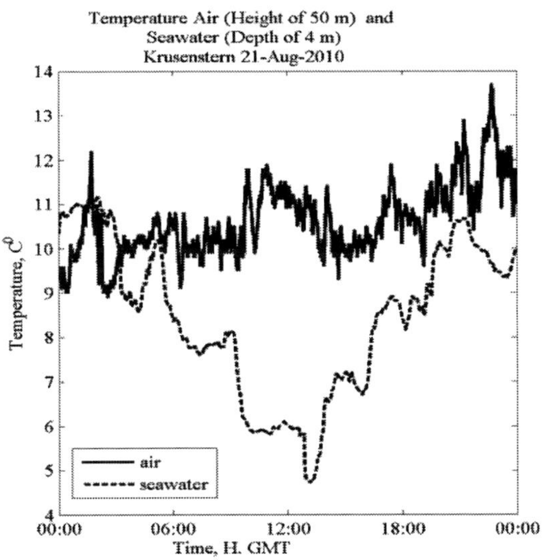

Fig. 4a (top) Chlorophyll-A concentration and bottom configuration

Fig. 4b (bottom) Temporal chances of water temperature (depth of 4m) and air (height of 50m)

Let us note that generally the same situation is observed on Krusenstern Bank: its boundaries see sharp chlorophyll-A increase at the level of 4 m.

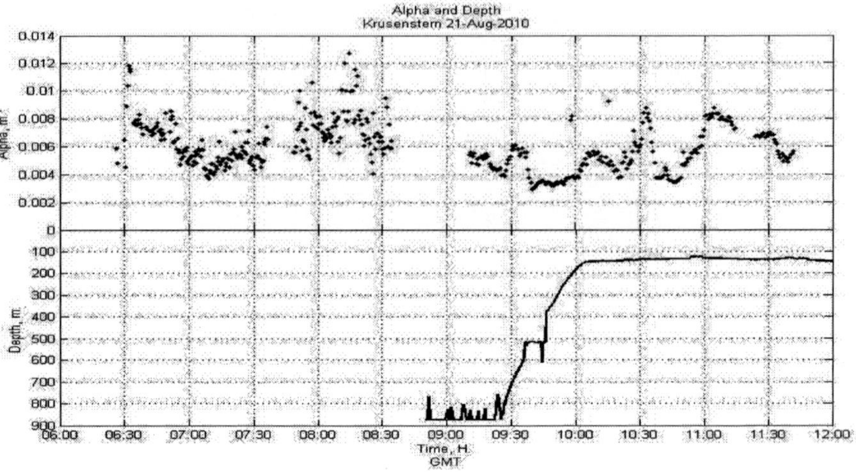

Fig.5 Temporal changes of PBL height (dots) above Kashevarov Bank

However, PBL height behaves differently from the way it does above Kashevarov Bank. It is especially clear on scattering diagrams based on temporal lines of extinction coefficient, chlorophyll-A concentration and humidity.

According to graphs, there is no definite correlation dependence between the extinction coefficient and chlorophyll-A concentration for Krusenstern Bank, its correlation coefficient being equal to 0.2. The scattering diagram of figure 6 singles out two relations. However, the correlation coefficient is low and one can not raise the question of any interrelation here.

The research has proved correlation dependence between chlorophyll-A concentration and aerosol content in the atmosphere to be either quite high (e.g., Kashevarov Bank) – the correlation coefficient being >0.6; or low (e.g., Krusenstern Bank) – the correlation coefficient being < 0.4. One can not assume with certainty that phytoplankton communities participate in aerosol field formation. This influence is extremely difficult to single out with indirect methods. One also should expect aerosol field formation processes caused by dimethylsulphide to appear in limited spatiotemporal areas.

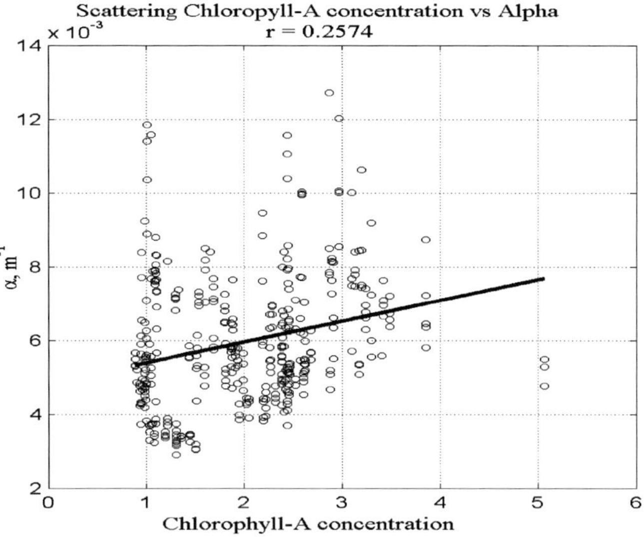

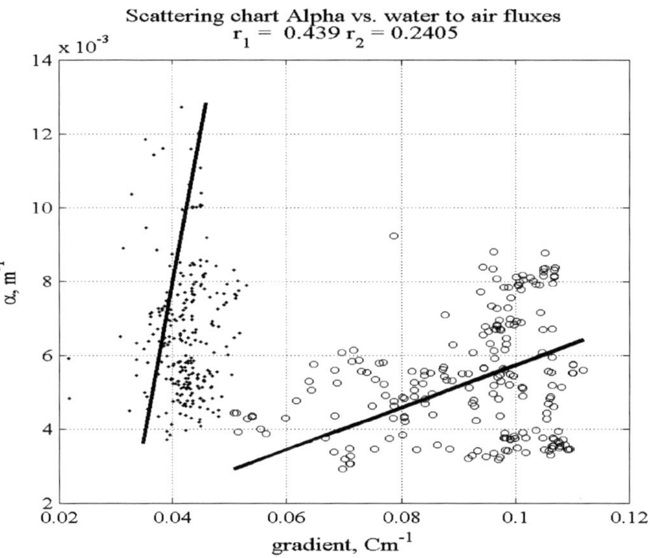

Fig.6a Scattering diagram of chlorophyll-A concentration and extinction coefficient. Fig.6b Scattering diagram of extinction coefficient and water-air temperatures gradient

Bifrequent Fluorometer Background Signal Research: Its Dynamics and Recording during Data Processing

Maxim Shinkarenko

This report researches the background signal recorded with a CCD camera, installed on a bifrequency lotic fluorometer. This equipment was used in the research expedition in August-September 2010. The route was as follows: Vladivostok, Korsakov, Raikoke Island, Matua Island, Kashevarov Bank, the Amur River Estuary, Krusenstern Bank, and Ekarma Island. During the expedition, the data were being collected along the entire route, and the spectrums obtained with a fluorometer were analyzed. The background signal was found to be occasionally changing. This could further lead to difficulties in signal spectrum processing, as the background level sometimes exceeded the signal level.

At the current processing stage, spectrums contained a large number of sharp single peaks (Fig.1). Such peaks were removed with a standard process of median filtration ***medfilt2*** for the Matrix Laboratory (MATLAB) programme. Application of this filtration to the three-dimensional bulk of data permitted to filter single peaks according to their measurement time. The ***medfilt2*** process consists in assigning a median of three surrounding points to each two-dimensional array.

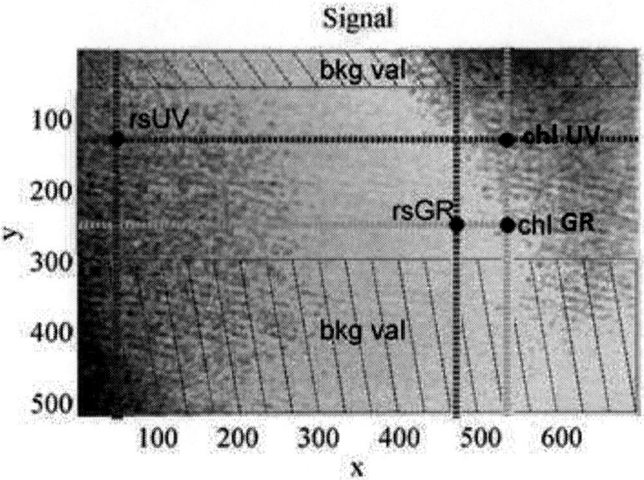

Fig.1 Three-dimensional illustration of background signal

This figure shows three-dimensional processing of the background signal (top view) after median filtration. Shading intensity shows background signal height in the studied area, while lines indicate areas in which the signal was recorded. Intersections of these lines possess intensity peaks of the ultraviolet (UV) area; the green area (GR); and chlorophyll, induced by the UV and the CHL GR. Figure 1 demonstrates clear division into areas of special importance, defined with UV and GR lines, as the fluorescence spectrum can be registered in this area, and quite large parts of it are marked with hatch lines. There is also a bkg_val area that does not and will not possess the fluorescence signal. This zone is extremely important in the process of background signal smoothing, as it permits to considerably smooth the signal. bkg_val is calculated as a compound of background signal intensity divided by the amount of pixels (n).

Bkg_val=

Let us make two-dimensional cuts out of areas we are interested in. As an example, let us take GR and UV areas. Figure 2 shows spectral cuts of these components depending on background signal intensity and bkg_val. These cuts clearly show a large number of peaks, that illustrate quite a considerable intensity range between single pixels. Therefore, various methods are necessary for approximation, e.g. spline approximation and bkg_val smoothing.

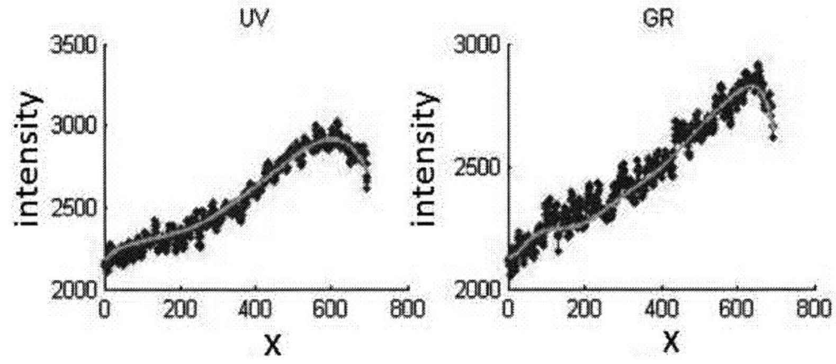

Fig.2 Two-dimensional cuts along X-axis of UV and GR depending on bkg_val.

In analyzing cuts exclusively according to background characteristics recorded at various time points, one can notice intensity changes. From time to time, various peaks can occur, which means the background signal is not stable, on the contrary, it is considerably changing in time. This can be observed, as during the background recording the fluorometer laser is off, and there is no temperature stabilization on either the CCD matrix or the electronic optical converter. All the above-said proves the vitality of recording the background as often as possible in order to more precisely assess the noise signal.

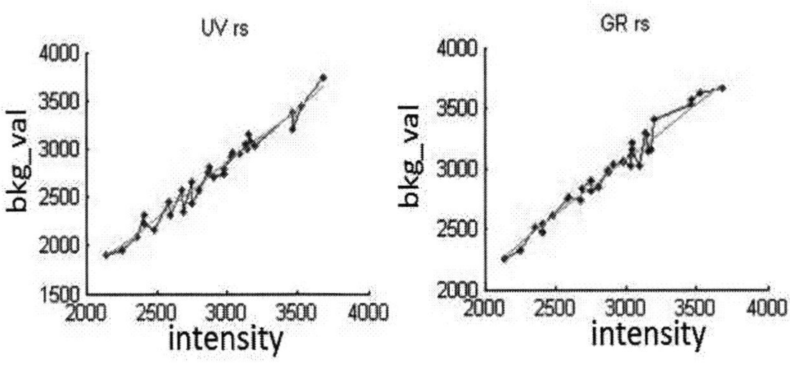

Fig.3 Two-dimensional cuts along Z-axis depending on bkg_val.

Figure 3 shows two-dimensional cuts of intensity and bkg_val parameters for combinational dispersion in the green area and the ultraviolet area.

The standard MATLAB–smooth function with a ***loess*** method (the method of locate square regression and a weighting window defined with a tricubic function) was used for noise smoothening. The ***loess*** method builds a square weighting polynomial in a slide weighting window, the window-centre value of which gives a smoothed value. The point of this chosen method is that there is no strong smoothening of intensive peaks, unlike in the moving average method.

Figure 4.1 presents a three-dimensional graph of noises in the UV area on the right, and the graph of noises in the chl area on the left. The figure distinctively shows that the higher the bkg_val level is, the higher noise intensity is, and the spectrum develops a more complicated structure. Figure 4.2 shows the same background signal after bkg_val smoothing.

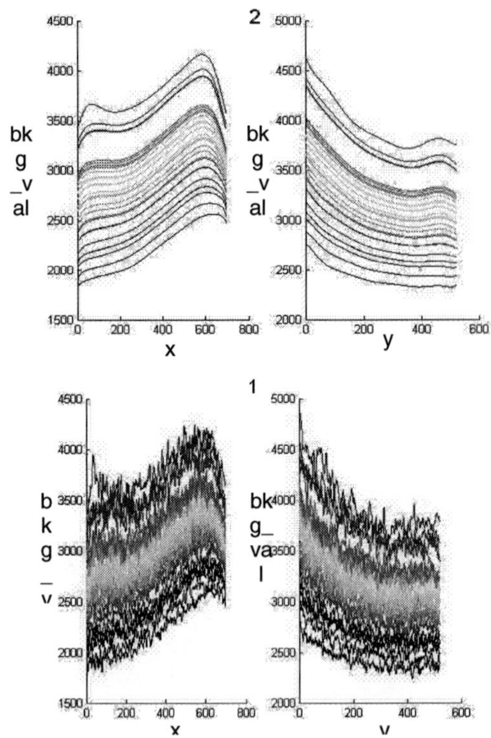

Fig.4. Background signal example: with noise and after bkg_val smoothing

Figure 5 illustrates a model of the background signal after three-dimensional spline approximation. A three-dimensional signal smoothed with median filtration and bkg_val is on the left. A two-dimensional cut of chlorophyll-A in the ultraviolet area is on the right.

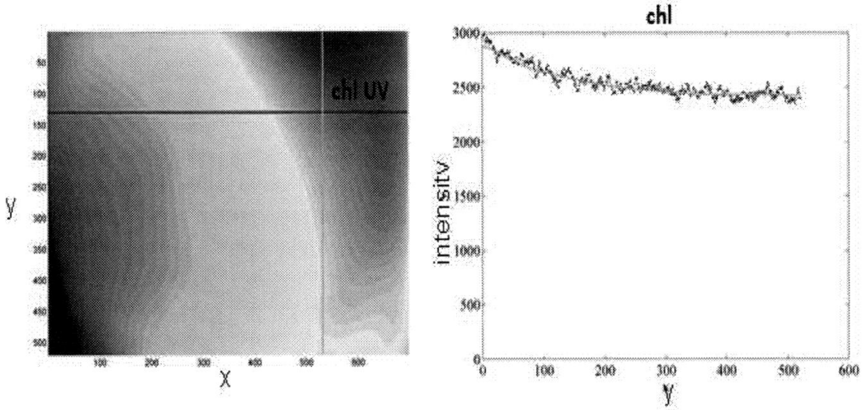

Рис. 5. Model presentation of background signal

Analyzed methods of obtained data processing are the most optimal for solving the problems we had set. The processing algorithm, shown above, brings the slightest distortions into the processed data. The use of the median filtration method helps smooth single fluctuations during processing of the obtained data without bringing additional inaccuracy into initial data. It should be noted that, in using the spectrum smoothing methods with various approximations and the filtration of single emissions, we can obtain numerical parameter values with high background accuracy, which can be excluded from the signal spectrum. This can boost the research of marine biologic productivity of phytoplankton communities, of anthropogenic influence and other global changes in the world ocean.

DEVELOPMENT OF INTERACTIVE SYSTEM FOR ENVIRONMENT MONITORING DATA ANALYSIS

ANDREY BRENEV

Research carried out aboard the training tall ship Nadezhda implies processing and storage of a large amount of various experimental data. In order to optimize data storage, exchange and analysis, as well as for educational purposes, the informational Internet portal lami.dvo.ru is being developed. An experimental database dealing with environment monitoring and accessible via web interface is to become a scientific foundation of the portal. Currently, the database is being locally developed and possesses an array of programmes necessary for data processing and analysis, which form the Interactive System for Data Analysis (ISDA). The System includes the following experimental data: atmosphere lidar sounding results, atmosphere meteorological sounding database, and information obtained from satellite and ship scanners of sea upwelling radiation, lotic measurements of sea water biooptical parameters, and information on abrupt climatic changes taking place in the Pacific Region. Currently, there are a number of well-known foreign analogues of such systems that provide a registered user with various data (satellite data, terrestrial data, etc.). Notable foreign analogues to be named: Ocean Color Web, Giovanni, ARL, weather station sites (Upper Air) and act-of-nature occurrence forecasting sites (in particular, RIAM CFORS, for dust storms). The key objective of the ISDA LAmI is a complex analysis of ocean and atmosphere parameters and their interaction in various special and temporal scopes. The System can be used for research of climatic change occurring on our planet.

The Interactive System for Data Analysis consists of the elements Figure 1 illustrates. The core ISDA element is a database with measurement tables and processing results. All tables are stored in one common database and are interconnected. Each table represents a separate laboratory and/or expedition experiment. One table that all other tables are referring to is the table of geographical reference. Any measurements entered in the database must refer to the lines of this table. The connection between experimental data tables and geographical reference tables is defined as one-to-many, i.e. for each line of the experimental data table; there is a sole date-and-time record in the geographical reference table.

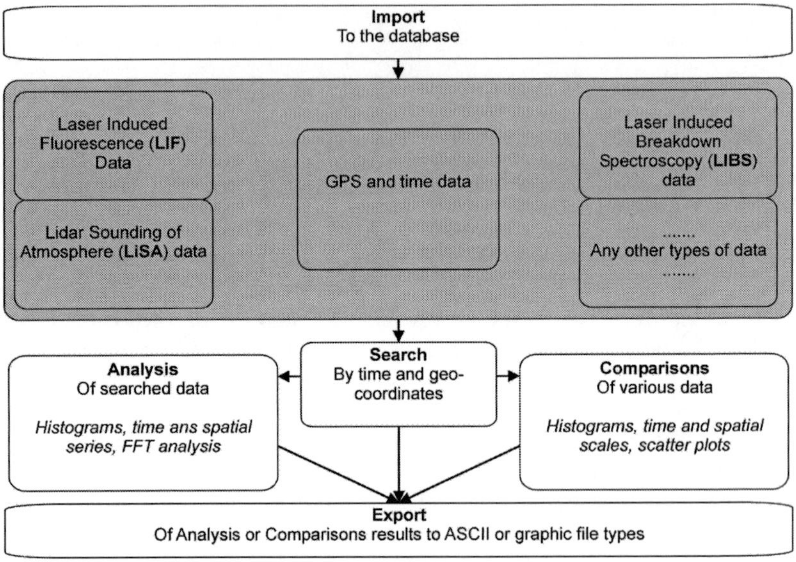

Fig.1 – Principle diagram of Interactive System for Data Analysis (ISDA) blocks

To avoid confusion in naming tables and their fields, the following guidelines for tables and their structuring are adopted.

Rule #1. Table names can contain no more than 4 symbols which fully characterize data entered into it and are unique in the database.

Rule #2. The symbol of '_' is used to specify the variable name or the table column.

Rule #3. If the length of the specifying element equals one symbol, the symbol of '_' is omitted.

Rule #4. If several identical equipment tools are working simultaneously, the equipment index number is added to the table name. This number is a two-digit number. In case it is less than “10”, preceding “0” is kept (e.g., lid01)

Rule #5. A unique name is assigned to each column.

Presently, the following tables have been created and filled in (Fig. 2):

asd – ASD radiometer measurements taken aboard the ship;
ctd – SBE-19plus immersion probe measurements;
dl – indicates that data were obtained via dataLogger (humidity, temperature, illumination);
fl – lotic fluorometric measurements (monofrequent and bifrequent stimulation);
gps – temporary and geographical reference data;
lic – an apparatus to measure water carbon dioxide concentrations (a lotic measuring instrument);
m – shows that the rotation method was used to obtain the data (water and air temperature, humidity, wind speed);
most – data obtained from the command bridge;
lid – lidar measurements (monofrequent sounding data, multifrequent sounding data, ozone sounding data);
sp – solar photometry.

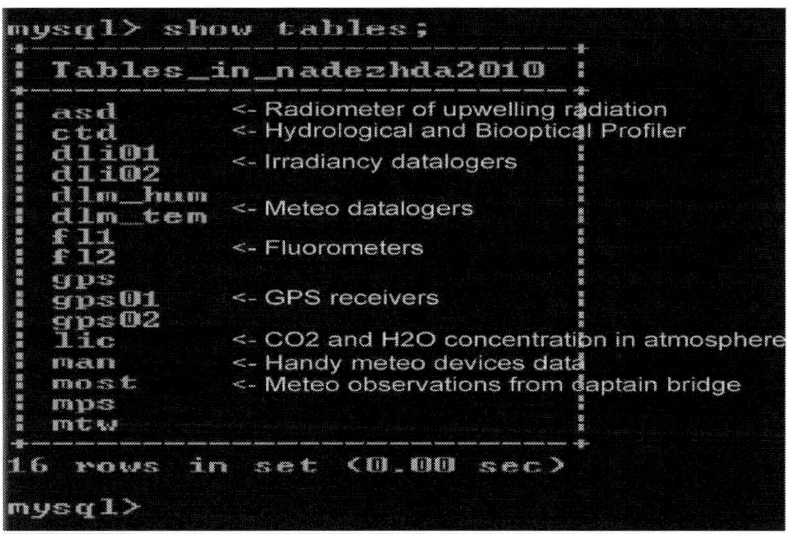

Fig. 2. Nadezhda_2010 database table list

Database structure:

Data are divided into the following processing levels in the database (Fig.3):

Level 0a – Source files with data stored on the hard drive. The absence of any processing is their particular feature.

Level 0b – The first processing has been carried out. Text data are transferred from measurement tables to corresponding databases. Binary files are sorted into special directories, while links to files with binary data and put into correspondent tables.

Level 1 – Geographical reference is assigned to all data in tables (each table line with measurements will have a link to the gps table).

Level 2,3,..N – Levels of data processing, containing processing results of the lower level data. E.g., level 2 can contain data obtained through data analysis of preceding levels (i.e., lower than level 2).

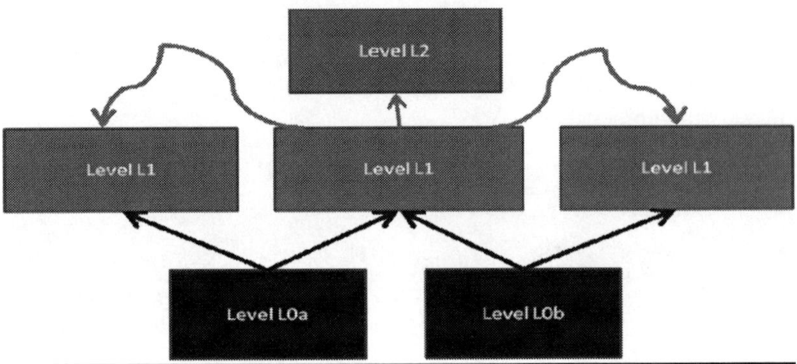

Fig. 3. Database structure

Technical characteristics:

The database is based on the mysql database management system. Numeric data (measurement tables) are stored directly in database tables (full transfer into the table). Large binary data are stored on the hard drive. The corresponding table simply indicates the way to the necessary file.

Web interface requirements:

Data visualization: either a single experiment, or as requested;

Possibility to export both "raw" data and processed results in various formats;

Request assignment for data processing (interactive mode).

Figure 4 displays an example of lidar data visualization.

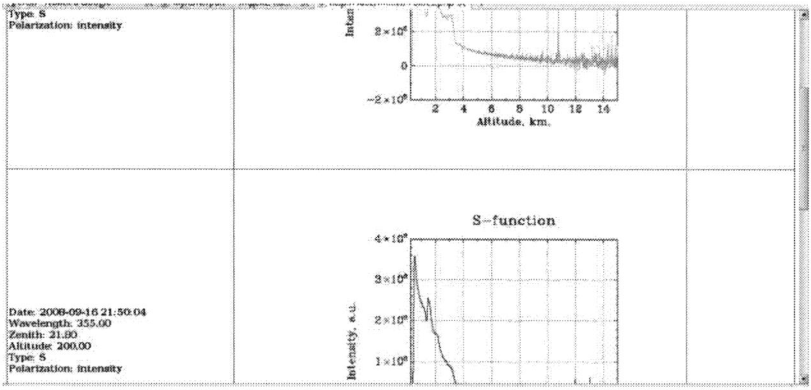

Fig. 4. Interactive System for Data Analysis interface

The entire data processing is carried out on the server. It includes source data processing, calculation request assignment, and calculation results visualization in a browser window. Algorithms aimed at optimal usage of experimental data and adapted for research in the transition zone.

So far, only FEB RAS network users are authorized to access the data on http://lami.dvo.ru/isad

LASER INDUCED FLUORESCENCE POSSIBILITIES TO DETECT METHANE ELEVATED CONCENTRATIONS IN THE SEAWATER

ILYA BUKIN AND DMITRY LYAKHOV

The issue of finding elevated concentrations of methane in seawater is extremely pertinent, not only in terms of finding deposits of gas hydrates, but also in terms of studies of the carbon cycle in the ocean. White S.N. et al and Hester K.S et al demonstrate the laser Raman spectroscopy application in order to study gas hydrates using remote-controlled robots. This paper discusses the usage possibility of the laser induced fluorescence method in order to detect seawater masses with elevated methane concentration, using a spectral apparatus developed for a remote operated vehicle.

The term "gas hydrates" defines crystalline compounds formed under certain conditions of temperature and pressure by the water and gas at low temperatures (close to zero degrees Celsius) and high pressures - at least 50 atmospheres. Gas hydrates resemble in appearance either compacted snow or dirty ice. Presently, the Russian gas hydrate resources are only found in underwater clusters, their amount being estimated as several trillions m^3. The main part of the underwater resources is located on the sea bottom surface at the depth of 300-400 m, in particular the deep shelf and continental slope, 500-600 m below the bottom. They easily disintegrate into the water and gas at higher temperatures, while the decay of one cubic meter of gas hydrate produces up to 160 cubic meters of gas. The issue with gas hydrates is that they are unstable under normal conditions in the laboratory or aboard the ship. Being lifted from the bottom, they decompose into the gas and water. Therefore, the study of gas hydrates should be carried out directly on the seabed. One of the underwater research methods is to implement various equipment installed on a variety of submersibles capable of reaching depths with gas hydrates on the seafloor. These submersibles are to carry out various works with gas hydrates: search; detection; visual inspection; sampling; internal properties measurement; decomposition processes study. Scientists, researchers and oil industry specialists from high-developed industrial countries pay special attention to gas hydrates studies. In the U.S., Monterey Bay Aquarium Research Institute organized research activities of gas hydrates at the depth of 850m, about 40 miles off the coast of Canada. Using laser spectroscopy ROV, Doc Ricketts conducted research of gas hydrates. It allowed obtaining specific data about the gas hydrates state in the place of their occurrence.

The Institute of Oceanology, Russian Academy of Sciences named after P.P. Shirshov carried out the research to create the test equipment in the submarine complex of remote control for marine gas hydrate deposits detection and assessment. The main search place of gas hydrates in Russia is concentrated in the Okhotsk Sea and Lake Baikal. First in the USSR searches for hydrate deposits were conducted on the Kuril Bank. Nowadays, gas hydrate resources in the Okhotsk Sea area estimated at 87 trillion cubic meters. In the area of the Eastern coast of Sakhalin Island, in the deep faults are located largest proven reserves of gas hydrates, more than 50 fields (Figure 1). Usually, scientific expeditions of the Far Eastern Sailing University aboard the sailing ship Nadezhda are carried out in the abovementioned areas. The combination of acoustic sounding of the upper ocean layer and laser induced fluorescence of seawater, carried out on the running vessel, enable the analysis of fluorescence spectra laser for such marine waters that are exposed to gas flares rising from the bottom of the Sea of Okhotsk. Such flares were sensed in the Kuril Strait during the expedition in 2001, and in the East coast waters of Sakhalin Island in 2001 and 2002.

Preliminary analysis of spectra fluorescence that were carried out with the pumped laser fluorimeter installed aboard the Nadezhda, revealed that in some cases there occured specific broadband peaks in the spectrum fluorescence seawater pumped laser with a wavelength of 532 nm. These peaks were at frequencies corresponding to the Raman scattering of methane molecules (Fig. 2). A possible explanation is that the gas plumes, methane, significantly increase the concentration of methane in water, and thus peaks of the Raman scattering of methane appear on the seawater fluorescence spectrum. Otherwise, most likely, the methane high content leads to the proliferation of specific bacterial fluorescence under laser irradiation and the changes in the spectrum.

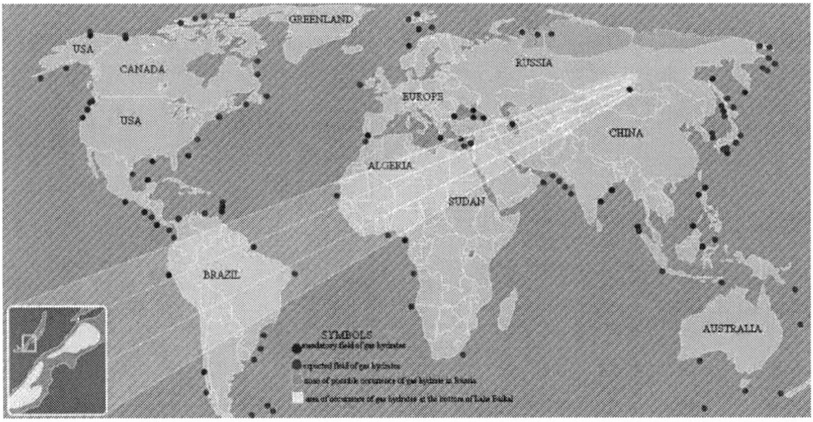

Figure 1. Deposits of gas hydrates. Red dots - fixed deposits. Blue dots - alleged deposits.

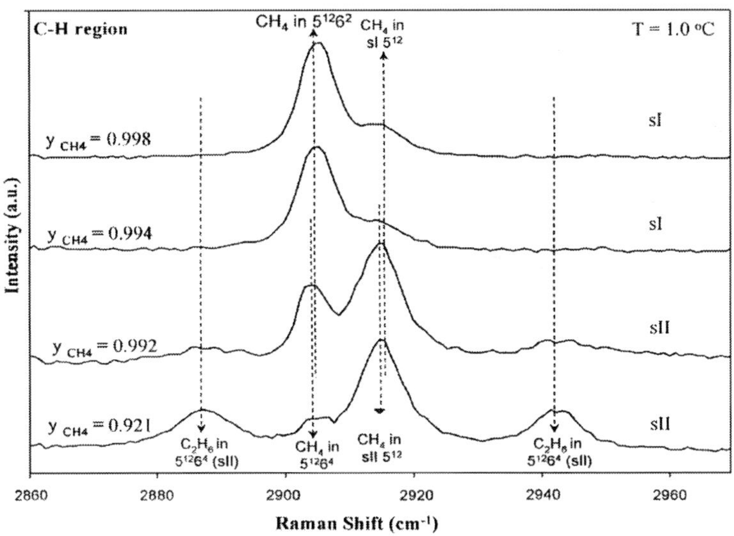

Figure 2. Raman scattering spectrum of methane [3].

Having calculated the mass, size, we plan to set an energy laser fluorimeter on a long-distance underwater device for remote measurements of fluorescence spectra of seawater at depth. It is shown that in order to ensure the ratio of signal to noise ratio, it is enough to use a continuous laser with a wavelength of 532 nm, average power of about 80 MW, an energy consumption continuous spectrum of the complex of about 100 watts, the necessary spectral resolution of the order of 1 nm, the recorded spectral range from 540 nm to 800 nm.

References

Hester K.S., Dunk R.M., White S.N., Brewer P.G., Peltzer E.T., Sloan E.D. Gas hydrate measurements at Hydrate Ridge using Raman spectroscopy. Geochimica et Cosmochimica Acta 71. 2007. 2947-2959.

Thomas John Hughes. Plug formation and dissociation of mixed gas hydrates and methane semi-clathrate hydrate stability. A thesis of Doctor of philosophy in chemical and process engineering. University of Canterbury.2008. p.42.

White S.N., Dunk R.M., Peltzer E.T., Freeman J.J., Brewer P.G.. In situ Raman analyses of deep-sea hydrothermal and cold seep system (Gorda Ridge and Hydrate Ridge). Geochemistry, Geophysics, Geosystems. V.7, # 5, 25 May 2006. p. 1-12.